Environmental Activities for Young People

GEORGE HENNINGS
Professor of Biology
Kean College of New Jersey

DOROTHY GRANT HENNINGS
Professor of Education
Kean College of New Jersey

CITATION PRESS
NEW YORK
1976

To our parents

**GEORGE AND CAROLINE HENNINGS
WILLIAM AND ETHEL GRANT**

Library of Congress Cataloging in
Publication Data

Hennings, George.
 Keep Earth clean, blue & green.
 Bibliography: p.
  1. Pollution—Study and teaching. 2. Environmental protection—Study and teaching. I. Hennings, Dorothy Grant, joint author. II. Title.
TD178.H46   1976   363.6   75-42336
ISBN 0–590–09403–3 (Paper)   ISBN 0–590–07460–1 (Hardcover)

Copyright © 1976 by George Hennings and Dorothy Grant Hennings. All rights reserved. Published by Citation Press, General Book Publishing Division, Scholastic Magazines, Inc. Editorial Office: 50 West 44th Street, New York, New York 10036.

Printed in the U.S.A.

Library of Congress Catalog Card Number: 75-42336

5 4 3 2 1   80 79 78 77 76

*Line drawings by George Hennings
Cover design by Honi Werner*

CONTENTS

PREFACE 1

1 TOWARD AN ALERT CITIZENRY
An Introduction 3

2 BRRR-RRR-RRR!
Activities with Noise Pollution 12
    INVESTIGATING VIBRATIONS 13
        Vibrating Objects 13
        Sounds Get Around 15
    INVESTIGATING NOISES IN THE ENVIRONMENT 17
        Which Noises Are Which? 17
        Quiet! I'm Trying to Think! 18
        Listen to the Noises 19
        The Relative Intensities of Sound 20
        Perceptions of Sounds 22
        Comparing Noises 24
        Sound Barriers and Muffled Sounds 25
        Traffic and Neighborhood Noises 27
        Noises in Homes and at School 29
        Noise Control Ordinances 30
    WRITING ABOUT NOISES 31
        Describe the Sound 31
        Sound Rhythms and Poetry 32
        Writing a Noise Control Ordinance 34

3 "THE AGE OF THE BIG THROWAWAY"
Activities with Solid Waste Disposal 36
    INVESTIGATING THE LITTER PROBLEM 37
        Litter, Litter Everywhere! 37
        Pictures Tell the Story 38
        Ban the Dog 40
        Johnny Horizon 41
        L Is for Litter 42

  FINDING OUT ABOUT REUSING AND RECYCLING 42
   "Think Solid Waste Before You Buy!" 42
   Don't Throw It Away! Reuse It! 44
   Recycling Bottles 45
   Recycling Iron and Steel 48
   Recycling Aluminum Cans 50
   Recycling Paper 51
   Paper to Paper 53
   Beware of the Hard Sell! 56
  INVESTIGATING DISPOSAL AND DECOMPOSITION 59
   What Shall We Do with Our Trash? 59
   Decomposition—Nature's Slow Solution 60
   Sewage—What Shall We Do with It? 62
   Halting Pollution 65

4 A THING OF BEAUTY OR DANGER
Activities with Water Pollution 67
  INVESTIGATING DRINKING WATER 68
   Particles in Water 68
   More Particles in Water 71
   Life in a Drop of Water 75
   Making Water Fit to Drink 75
  INVESTIGATING WATER ENVIRONMENTS 79
   Looking at Water in Nature 79
   How Deep Can We See? 82
   How Fast the Water Travels 84
   Oxygen, Temperature, and Life 86
   Life in a Healthy Stream 89
   Life in Polluted Streams 92
   Life in Lakes and Ponds 93
   Chemicals in Our Water 97
   Acid or Alkaline? 100
   And How Do We Use Our Water Environment? 101
  LEARNING MORE ABOUT WATER POLLUTION 102
   Getting the Facts 102
   Developing Case Study Reports 104
   To Eat Clams or Not to Eat Clams? 107
   This Is Our Country 109

5 DENUDING THE FIELDS
Activities with Soil and Pavement 110
  INVESTIGATING SOIL-WATER RELATIONSHIPS 111
   The Water Cycle 111
   Will Water Penetrate? 114

Measuring Raindrop Splash Power   117
  Is There a Crust?   119
  Runoff and Erosion   119
  INVESTIGATING PAVEMENT-WATER RELATIONSHIPS   123
    Paving the World   123
    Pavement vs. Grass   125
  MAINTAINING SOIL PRODUCTIVITY   126
    Looking for Soil Profiles   126
    Interpreting Soil Profiles   127
    Soil Testing—Acid or Alkaline?   133
    Soil Study—Other Minerals   134

6 "THE THREAT OF SUFFOCATION"
Activities with Air Pollution   138
  INVESTIGATING AIR   139
    Air Is Really Real?   139
    What Is Air?   143
  INVESTIGATING CONTAMINANTS IN THE AIR   145
    How We Contaminate the Air   145
    Seeing Particles in the Air   149
    Vacuuming the Air   149
    Sampling Particulates with Sticky Traps   150
    Sampling with Dustfall Catchers   154
    Checking Dirt in City Air   155
    Sampling Snow   156
    Our Community   157
    Air Pollution Mapping   158
    Smog Weather Watch   160
    Winds and Pollution   162
    An Inversion Lid on the PAN   164
  INVESTIGATING EFFECTS OF AIR CONTAMINANTS   169
    Danger to Man!   169
    And Plants Too!   172
    Nature's Threats   174
    Damage to Materials   177
    Documenting Air Pollution   179
  CLEANING UP THE AIR   181
    Clean Air?   181
    Don't Burn the Leaves   182
    A Clean Stack   183
    Individual Versus Community Rights   184
    A Job for You   185
    A Song for You   186

contents   v

7 ENERGY CHALLENGES
Activities with Energy Problems 187
INVESTIGATING ENERGY SOURCES 188
Energy from Coal 188
Energy from Oil and Gas 192
Petroleum from Sands and Shales 196
Nuclear Power 197
Heat from the Earth 200
Solar Power 201
Hydroelectric Power 204
Power from the Wind and the Tides 206
Energy Stored in Plants 208
Garbage, the Cinderella Fuel 209
ANALYZING ASPECTS OF ENERGY PROBLEMS 211
Doing the World's Work 211
Analyzing the Issues 214
Conserving 216
Investigating Local Energy Consumption 217
It's Controversial! 219
Meet the Press 220
Working and Writing for Change 221
The Energy Song 224

8 A PARTING THOUGHT 225

ANNOTATED RESOURCE BIBLIOGRAPHY 227
TEACHER REFERENCES 227
BOOKS FOR YOUNG PEOPLE 231
FILMS FOR CLASSROOM USE 236

PREFACE

Keep Earth Clean, Blue and Green is a source book for elementary and junior high school teachers. It provides numerous activities for involving young people in interdisciplinary environmental investigations. Investigative procedures are described—often step by step—and some background information is supplied in instances where supporting concepts are complex.

In selecting and developing activities, we have tried to include a variety that range in difficulty. Some activities are relatively simple for use with young children who have had little experience conducting investigations. Others require a greater degree of experience and maturity. We have intentionally not assigned grade level designations to most activities, believing that students in specific grades differ to such an extent that level designations are inappropriate. The teacher who knows his or her students is best able to determine whether a particular investigation will be practical. Then too some of the more complex activities can be simplified by the teacher so that they can be used in the lower elementary grades.

We have tried also to include different kinds of activities. Some are investigations requiring science laboratory procedures or systematic surveying techniques. Others are ideas for library research and for sharing findings with the class. Still others are art-, poetry-, or music-related experiences.

When planning any of the investigations with their classes, teachers will find that adjustments will probably be necessary. This we anticipate since availability of materials and local environmental conditions differ widely and since young people will be involved in planning the actual investigations. In this respect, the activities in *Keep Earth Clean, Blue and Green* serve as a guide rather than a directive.

We would like to extend our appreciation to all the companies, institutes, and public agencies that have sent us information and brochures. To Mary L. Allison, our editor, we say a special "Thank you" for ideas and editorial assistance. To colleagues with whom we discussed activities, we also add a "Thank you."

G.H.
D.G.H.

CHAPTER 1

TOWARD AN ALERT CITIZENRY
An Introduction

It is mandatory that every young scientist, and indeed every educated person, acquaint himself with at least the overall environmental processes and conditions that make possible the very survival, not to mention the prospering, of individual organisms such as ourselves. In a democracy it is not sufficient just to have a few trained persons who understand what it's all about; there must also be an alert citizenry to insist that knowledge, research, and action are properly integrated.

Eugene P. Odum, Ecology (Holt, Rinehart and Winston, 1963, p. 1)

NEWS ITEM! July 1969. "One of the nation's most polluted streams, Ohio's Cuyahoga, became so covered with oil and debris that . . . the river caught fire here in Cleveland's factory area, damaging two railroad bridges. Along this six-mile stretch, before emptying into Lake Erie, the river receives the wastes of steel mills, chemical and meat-rendering plants, and other industries. Just upstream, Cleveland and Akron discharge inadequately treated sewage. And from hinterland farms drain phosphate- and nitrate-rich fertilizers and poisonous pesticides." (*National Geographic*, December 1970)

NEWS ITEM! January 28, 1972. An East Coast power and light company "shut down its nuclear power plant in Oyster Creek at Forked River, Lacey Township, Ocean County. The plant uses water from the south branch of the river to cool its condensers. During the cooling process the temperature of the river water is raised; the water then is discharged into Oyster Creek. The power company continued to pump water from Forked River to Oyster Creek during the shutdown. The sudden introduction of the cold (under 40 degrees) water from Forked River into the warm (well over 50 degrees) water of Oyster Creek dramatically changed the environment of menhaden fish inhabiting the waters with the result that hundreds of thousands were thermally shocked and died." (*New Jersey Environmental Times*, November 1973)

NEWS ITEM! January 11, 1974. "Europe has been hard hit by the energy crisis. More dependent on imports of Arab oil, most European governments are now facing up to the real prospect of a drop in living standards next year and a continuing heavy burden on their balance of payments produced by the increased oil prices.

"In Britain the situation has been made a great deal worse by bans on overtime working by coal miners and electricity power engineers. . . ." (*Science*, January 11, 1974)

NEWS ITEM! February 3, 1974. "The famous bronze horses over the main portal of St. Mark's Basilica in Venice are suffering from 'bronze cancer' caused by pollution, and authorities are not sure how to save them . . . .

"Pollution in Venice is different from that in other Italian and southern European cities. While there is little engine exhaust because there are no cars in the city of canals, the air is laden with salty humidity and sulfurous pollutants from refineries and petrochemical industries on the nearby mainland and from home and hotel heating equipment." (*New York Times*, Feb. 3, 1974)

NEWS ITEM! January 20, 1975. "Oil *Shokku* for Japan. To

keep its economy healthy, Japan must receive one fully loaded supertanker every hour of every day. It must also get oil from the Middle East via the shortest route and then provide vast storage facilities for the vital fuel—all without undue environmental risk. Until recently, the Japanese were confident that they could transport and store their oil safely and efficiently. Now two serious oil spills have caused *shokku* (shock) and raised grave doubts on both counts." (*Time*, January 22, 1975)

NEWS ITEM! September 14, 1975. "Climatic Changes by Aerosols in Atmosphere Feared. Although attention in the debate regarding the role of fluorocarbon aerosol sprays has hitherto focused on public health questions, it has now been proposed that they could alter world climates." (*New York Times*, Sept. 14, 1975)

News items such as these, which zero in on pollution and energy problems, are becoming more and more common. Their prevalence indicates a mounting belief that "something must be done" about the way we are fouling our environment and wasting our resources. This "something" may well be intensive involvement of young people in problems cause by a population living on a planet with limited resources while rapidly increasing in number and expectations. Young people can investigate ecological relationships, survey current conditions, and play an active part in restoring the environment and in seeing that natural resources are consumed wisely. As students probe problems, they may gain a working understanding of and a respect for the interrelatedness of living things with their biotic and physical environments. In short, they may become functioning members of an alert citizenry.

Because of the complexity of environmental concerns, present day social conditions, and the kinds of learning goals believed to be fundamental, designers of environmental programs for young people may find it helpful to consider the following ideas:

Environmental study is interdisciplinary, utilizing concepts and approaches from the biological and physical sciences, the social sciences, mathematics, language, and the arts. Investigators of an environmental problem may employ the experimental techniques of the science laboratory, setting up controlled experiments and systematically recording data. They may utilize the surveying techniques of the social scientist, devising questionnaires to collect data and employing mathematical procedures to analyze that data. In reporting investigators must express hypotheses and conclusions in clear, logically organized prose and in so doing draw upon language skills. Or investigators may record data graphically to express what was observed. And, having worked with facets of the environment, they may come to feel deeply about what they have seen and express those feelings in poetry.

Concepts basic to environmental study must similarly be drawn from a number of disciplines. The scientist contributes concepts about the ways matter and energy behave and about relationships between living things and the environment. The economist contributes notions about supply and demand; the geographer about relationships among natural resources, topography, and population; the political scientist and sociologist about relationships between nations that have and have not and about relationships among peoples. Solutions to environmental problems usually require consideration of concepts from more than one of these disciplines. Otherwise, a narrow focus may result in partial answers.

Environmental problems are complex. When investigators look at an environmental problem such as air, water, or noise pollution or the energy shortage, they are actually encountering problems that spill over into other environmental areas. For example, the problem of disposing of heat from a nuclear reactor as in the Forked River-Oyster Creek incident relates to energy generation, water pollution, waste disposal, and even air pollution problems. Although for the purpose of organizing *Keep Earth Clean, Blue and Green* we selected a topical approach—each topic relates closely to other topics. Again, solutions to problems must involve consideration of many related issues to avoid partial answers.

In the past simplistic answers have been proposed based on only one factor or a few, e.g., ban strip mining of coal because such mining leaves ugly scars on the earth. The simplistic approach has also been a hysterical or maudlin one clothed in such phraseology as "Save the birds!" "Ban nuclear power plants!" or "No more dumping!" Environmental study must go beyond the superficial and focus on complex interrelationships.

Principles of ecology can serve as unifying threads that pull diverse elements together. Ecology is the study of organisms at home—their home; it encompasses the dynamic—sometimes subtle, sometimes clearly apparent—interactions among organisms and between organisms and their biotic and physical environments. Basic principles include:

☐ Each organism functions within the constraining influences of its environment. For example, a trout may live in a clear stream containing an adequate food supply, but if the water temperature becomes too high, the trout dies. A green plant has good soil, sufficient moisture, and adequate sunlight, but it cannot produce seed if its flowers are eaten by an insect. Carbon dioxide is normally in the environment in small amounts but becomes toxic at certain concentrations. Thus near Key West, Florida, two aquanauts died because carbon dioxide built up in their trapped diving craft. In each of these cases, an environmental factor constrained the organism and determined whether life continued.

☐ Reserves of materials and energy on planet earth are limited. Energy cannot be recycled; in every energy transformation, some energy is lost into space. To build new reserves, energy must be captured from the sun. In contrast, materials—minerals, metals, gases—can be recycled and used over and over again.
☐ Through photosynthesis green plants store solar energy in energy-rich compounds that they and other organisms can use. All other organisms, with the exception of a few chemosynthetic bacteria, cannot manufacture food and must depend on green plants. When humans do things that affect the growth and reproduction of plants—our primary producers—we are indirectly affecting ourselves and other organisms in the food chain.

Environmental principles apply to life in urban, suburban, and rural areas. Environmental problems are not limited to farmers concerned about soil erosion on lands under their cultivation. Environmental problems are becoming even more pressing in urban areas where the buildup of noise and of contaminants in the air and water supply increases daily and in which there is often not enough energy to go around. For this reason, environmental study is as important in urban and suburban areas as it is in rural areas.

Environmental problems are found close to home. It is not always necessary to investigate decreasing energy supplies or the destruction of the environment at some location far from home. Problems probably abound in the home neighborhood in the form of litter discarded in a vacant lot, noises from a nearby highway, emissions from the smokestack of a local public utility, or erosion from an area recently cleared for construction. Investigation of these problems is generally more meaningful to young people than a study of an esoteric problem occurring halfway across the continent.

A variety of goals can be achieved through environmental study. Goals of environmental study include thinking skills such as the ability to distinguish fact from opinion, to categorize, to generalize, to hypothesize, to apply fundamental scientific, social, and mathematical concepts to new situations, and to arrive at decisions based on an analysis of data. Goals include actual changes in the way students behave toward the environment, a deepened feeling for the environment, knowledge of the kinds of factors that must be considered when making decisions about the environment, and an ability to work for change using all the social processes available to citizens. An ancillary goal is the ability to express oneself on issues clearly in written and oral form.

Achievement of goals mandates active involvement of students. Listening to a teacher talk about the environment probably will have little effect on students' behavior, their commitment to environmental preservation, or their ability to think about problems. To become skilled at distinguishing fact from opinion, learners must have opportunities to distinguish facts from opinions; to become skilled at applying concepts to the analysis of new problems, learners must have ample opportunities to apply concepts. Firsthand involvement in problems is the name of the game.

Development of environmental awareness should begin early. Young children can begin to develop an understanding of and a respect for their environment. By looking for signs of pollution in their own neighborhoods, by identifying examples of excessive waste in classroom and home, and by thinking about ways they themselves contribute to environmental destruction, youngsters can break habits of wastefulness and thoughtlessness before those habits have become deeply ingrained.

Both girls and boys should be active investigators. Girls and boys should become experimenters, doers, recorders, and thinkers. Teachers must beware of the unfairness and irrationality of assigning girls the jobs of recording information and writing up investigations and assigning boys the jobs of actually carrying out the experiments. Don't make the girls stand by as the boys light the Bunsen burners, till a piece of land for cultivation, or ask questions.

Investigations can be conducted within several instructional frameworks. Organizational frameworks include small group activity, individual activity, total class investigation, and teacher-led discussions and demonstrations. For example, a class can work as a whole to investigate a litter problem along a local river bank, or groups can investigate facets of a problem and pool data on which they then formulate generalizations and hypotheses. Or individual students may select investigations that particularly interest them and, report their findings later to other individuals or groups in the class. Or, when investigations require the handling of potentially dangerous chemicals or materials, a demonstration by the teacher is a viable instructional framework with students observing, analyzing results, and hypothesizing. Ideally the choice of framework appropriate to a particular activity depends on the nature of the investigation and the specific goals sought.

Environmental study in schools as a way of doing something about improving our environment and stopping the waste of resources requires a financial commitment by the community. To put it in specific terms—discarded materials such as margarine tubs can be used in some investigations to collect materials or hold samples, but at other times more specialized equipment is essential, e.g., science glassware, propane torches, acid test papers, ring stands, or dip nets. Adequate funds must be available to purchase this type of equipment.

Resource materials used in investigating the environment range from the actual to the representational and the linguistic. Realia abound—lumps of coal, water samples, sound-producing instruments, soil, water-living organisms, even pieces of litter gathered from a roadside, to name just a few. Actual objects such as these can provide a concrete base for concept development.

A variety of representational materials are available or can be made by teachers and students—models: steam-driven turbines, geysers, volcanoes, or oil rigs; visuals: pictures, diagrams, charts, slides, and transparencies; audiovisuals: films, sound filmstrips, TV, and videotapes.

Linguistic materials include a wealth of informational books, storybooks, brochures and reports distributed by industries and public agencies, newspapers, and magazines. These are available in written form and sometimes on tapes and records. Some are free or relatively inexpensive. Specific examples are identified in the resource bibliography at the end of this book.

An environmental education classroom is not enclosed within four walls. It extends beyond the walls of the school to the vacant lot next door, the corner supermarket, a local stream or pond, the sidewalk in front of the school, or a nearby construction site. Student investigators should use their community as an environmental science laboratory for firsthand observations, data gathering, and action. They can return to their indoor laboratory to check references, organize data, draw conclusions, and investigate related problems.

The goal sought by extending the walls of the classroom is basic to all environmental education activity—that students perceive themselves as part of the environment, not its controlling force, and know that they must live in harmony with nature.

CHAPTER 2

BRRR-RRR-RRR!
Activities with Noise Pollution

My ears are shaken with an incessant whir. The air-drill chatters, the riveter palpitates. "Brrr" goes the world; "Brrrr-rrr-rrr!"

Morris Bishop, *Lines Written in a Moment of Vibrant Ill-health*

Do you have acoustical tile on your classroom ceiling? You know how much noisier the room would be if classroom sounds bounced back down on you after reflecting from a smooth plaster ceiling. Perhaps you also are lucky enough to be working in a carpeted room. How much easier it is to have several activities going on simultaneously when noises from moving feet and moving furniture are absorbed!

Have you ever said, "There ought to be a law against dogs barking at night, utilities digging with pneumatic drills or highway trucks that make your body vibrate?" Have you ever thought, "Would trees help?" or, "Why aren't people more sensitive to how their noises carry?" or, "Why was the transistor radio ever invented?"

Rather than listening to teachers tell about the sources of noise, how noise is carried and muffled, or about how it affects our lives, young people can have direct experiences with and make comparative observations of noises in their environment. This way they will gain a personal perspective on the problems of noise pollution and a heightened awareness of the effects of noise. This chapter suggests specific activities to involve students in sound phenomena that directly affect them.

INVESTIGATING VIBRATIONS

Vibrating Objects

Introduce students to problems of noise pollution by helping them to understand the nature of sound—how sounds are produced and how they travel. Begin with a few simple laboratory experiences in which three- or four-person teams analyze vibrating objects and the sounds produced.

Make available different kinds of rhythm instruments to strike and observe: tamborines, sticks, bells, and clackers; a tuning fork and a hammer made from a small rod pushed into the hole of a rubber stopper; a vessel such as a large margarine tub filled with water into which to plunge vibrating objects; thread and cotton wool; a Ping-Pong ball on a string; a glass tumbler, a piece of stretchable material such as plastic wrap to make a drumhead, a rubber band, and some iron filings or sand; a ruler; test tubes; other rubber bands; soda straws.

Also prepare a things-to-investigate sheet that suggests:

THINGS TO INVESTIGATE

Today we are going to investigate how sounds are made. The question is: What must happen in order for a sound to be produced? Do some or all of the following activities or any others that you invent to find an answer or possible answers. Be ready to describe at least one activity your team carried out and to explain what happened.

1. In sequence, sound each rhythm instrument displayed, observe how each produces its sound, and talk about how the sound is produced.

2. As you talk, with your fingertips, lightly touch your vocal cords in the middle front of your neck. What do you feel happening?

3. Gently strike the tuning fork with the stopper-type hammer and touch your fingertip to the ringing fork. What do you feel happening? Now strike the fork again, this

time sticking the prongs into the water that you have poured into a margarine tub. What do you see happening?

4. Make a ball from a piece of cotton; tie a string around it. Now hang the cotton ball against the sounding tuning fork. Repeat the experiment, this time hanging a Ping-Pong ball against the fork. What happens?

5. Make a drum from the tumbler, the plastic wrap, and the rubber band. Strike the surface with your finger. What happens to the surface when it is struck? Repeat the test, this time first sprinkling the drum surface with a little sand or iron filings. What do you see happening?

6. Place a ruler on a desk with one end projecting over the edge. Holding one end tightly on the desk top, bend the other end downward and then release. What happens?

7. Try to produce a sound with the test tube, the rubber band, and the soda straw. Once you have produced the sound, do something to change the pitch—the highness or lowness of the tone you produce; do something to change the amplitude—the loudness of the sound.

8. Once you have produced sounds with the rhythm instruments, try to do something to change the amplitude and the pitch of the sound produced. Were there situations in which you could not vary the pitch? the amplitude?

9. Invent your own musical instrument using the principles of sound production you have discovered.

After students conduct several investigations, hold a reporting session. Investigators can share their findings and inventions with the total class. Conclusions to be developed include: *1*) sounds are made by vibrating objects or air columns; *2*) sound vibrations can be made by pounding, plucking, and blowing on objects, and *3*) sounds differ in pitch and amplitude. As students demonstrate sounds and instruments, ask, "What is vibrating to produce sound waves?"

Young children can work as a class to study how sounds are

produced. After youngsters have enjoyed a musical rhythm band session, they can be encouraged to handle the vibrating instruments to feel the vibrations; they can then talk about what they think is happening.

Sounds Get Around

To demonstrate to students that sound cannot travel through a vacuum, suspend an electric bell in a jar from which the air can be evacuated with a vacuum pump. To do this efficiently, use a wide-neck jar into which you can put a two-hole stopper. Connect the vacuum pump to a piece of glass tubing extending through one hole of the stopper. Extend insulated copper wires attached to the bell through the other hole, using modeling clay to make an airtight fit. Connect the wire to dry cell batteries and switch as shown in the diagram. As the air is removed from the bottle, students will see the bell hammer continue to strike but will hear the sound diminish until it can no longer be heard at all. Ask students to hypothesize what will happen when air reenters and give reasons to support their hypotheses.

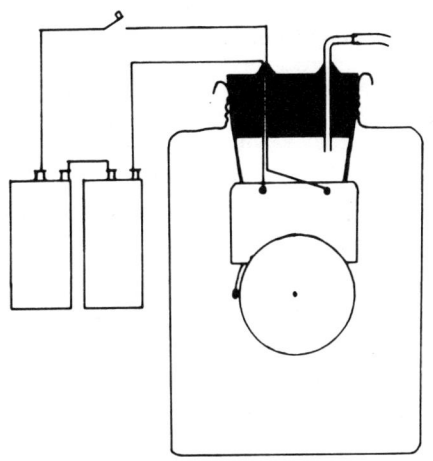

BELL IN VACUUM JAR

Students can go on to investigate the transmission of sounds through different materials. Make available string, round oatmeal boxes, tuning forks, corks, an aquarium with water, marbles, a slinky, a flat pan, and chart paper. Prepare a guide of suggested investigations from which pairs of students can choose. Include activities such as:

Hold an ear against a desk while your partner scratches her or his fingernail against the opposite end of the desk. Try the same experiment by holding an ear against one end of the chalkboard while your partner rubs an eraser across the board about a half meter away. What happens to the sound as the point source—the moving eraser—moves farther away—1 meter, 2 meters, 3 meters?

Thread each end of a two-meter string through the bottoms of two round oatmeal boxes; tie a knot on each end to hold the ends inside the boxes. Keeping the string taut, hold one box against your ear while your partner whispers into the other box. How loudly must the talker speak to be heard through the string telephone? What happens if you touch the vibrating string?

Float a cork on the water in the aquarium. Hold one ear against the glass of the aquarium, and cover the other ear with your hand while your partner strikes a tuning fork and touches the end of the floating cork. What path does the sound follow to reach your ear?

Make a model to see how sound travels through solids, liquids, and gases. Push two desks together so there is just a crack separating them. Set six or seven marbles against one another in the crack, and roll another marble along the crack so it will strike the first marble in the line. What happens? This shows how sound energy is transmitted through materials.

Extend a slinky—the loose metal coil—across the floor. Pinch it near one end. What happens? Now hold one end while your partner jiggles the other. What happens? This is comparable to the way sound energy travels through materials.

Drop a pebble into the flat pan filled with water. Watch the wave fronts ripple outward in larger and larger circles. Are the outer ripples weaker or stronger than the inner ripples? If these were sound waves, how would the distance from the point source affect the sound heard?

Draw a large-scale diagram showing a method of insulating against sound. Would air, wood, or water make good insulators against the transmission of sound?

After partners have completed at least one of the activities, they report to the total class. Conclusions to be developed at this point include: 1) sound travels in waves through liquids, solids, and gases; 2) sound is not transmitted in a vacuum so a vacuum can make a good sound insulator; and 3) sound is a form of energy.

Most high schools and colleges own an oscilloscope that visually depicts waves on a screen; that is an excellent way of explaining the different waves produced by sounds of varying pitch and loudness. Differences in frequency and amplitude of waves are clearly visible on the oscilloscope screen. See if a physics teacher from your local high school or college will demonstrate an oscilloscope to your class. Ask students who play musical instruments to bring their instruments to class so that the wave patterns they produce can be compared.

INVESTIGATING NOISES IN THE ENVIRONMENT

Which Noises Are Which?

The work of Hilda Taba suggests that young people should be given opportunities to categorize phenomena to develop their ability to handle contrasts and comparisons. Noises differ in source, pitch, loudness, and distance at which emitted; for this reason students will find it rather easy to invent their own categorizing systems and use them to classify noises in their environment.

Begin with a brainstorming activity. One student can serve as a recorder at the chalkboard. Others in turn call out noises they hear in the room, noises they hear coming from the street, or noises they hear at home. The recorder writes everything on the board. Then the teacher can take over as discussion leader by asking: Which of these noises are very much the same? How are they the same? What other noises are similar to those already singled out? Which noises are very much different from the others? A simple classification system can result, such as:

Man-made noises ——— Natural noises
Pleasant sounds ——— Unpleasant sounds
High sounds ——————— Low sounds

Once students have developed a simple classification system for noises, they can return on another day to their lists and see if they can organize the noises into another set of categories. From this activity students can perceive that it is possible to group any set of phenomena into a number of different classes. They also become aware of the multitude of sounds bombarding them.

Quiet! I'm Trying to Think!

Through a simple bulletin board activity, lower-grade children can gain increased awareness of the multitude of noise sources in their environment, skill in grouping related phenomena, and skill in writing sentences that describe noises.

Attach a card holder—a shoe box will do—to a bulletin board captioned "Quiet! I'm Trying to Think!" Also attach a flow pen. Divide the bulletin board in half and label one side Sounds of Nature, the other Sounds of Machines. During independent study times, youngsters may take a card, print on it the name of a machine or a natural source that emits a sound, and write one sentence that describes the sound. A rule in bulletin board making is to check spelling before mounting written work, so dictionary checking is part of this activity. After checking the spelling of their sentences, students select the category to which the sound source belongs and pin their cards to the board.

To turn the activity into a guessing game, youngsters can fold down part of the card so only the description of the noise is visible. Other class members must guess the source.

Listen to the Noises

Once young people have absorbed the notion that sounds—like other phenomena—can be categorized, they can use their self-devised classification systems as recording guides for observing traffic noises. Because schools generally do not have the instruments necessary for highly sophisticated analyses of environmental noise, student classification systems will probably lack sophistication and be dependent on subjective measures.

A simple analytical scheme that students may develop is based on loudness categories: ear-shattering, annoying, acceptable, barely perceptible. Students can stand in the schoolyard to observe traffic passing the building; they class each passing vehicle in one of the four categories. A dittoed sheet will help those who have limited experience observing and recording observations:

A GUIDE TO ANALYZE THE LOUDNESS
OF TRAFFIC NOISES

**DESCRIPTION
OF VEHICLE RATING**

1 _____ ear-shattering annoying acceptable barely perceptible
2 _____ ear-shattering annoying acceptable barely perceptible
3 _____ ear-shattering annoying acceptable barely perceptible
4 _____ ear-shattering annoying acceptable barely perceptible

---

Students merely place a descriptive word such as truck, car, or motorcycle in the blank and circle what they feel is the category of its sound. If traffic is very heavy, the class can divide into four teams, with each team responsible for evaluating the noise emitted by every fourth vehicle.

When students return to their classroom, a small work group can compile the data gathered by all the students and develop a composite analysis by averaging the classifications. Thus if twenty students called the sound emitted by vehicle 3—a truck—ear-shattering, seven classed the sound as annoying, and two classed the sound as acceptable, the composite classification would be ear-shattering.

Follow-up discussion can include: Why do some vehicles emit so much more noise than other vehicles? What can be done to make vehicles quieter?

The Relative Intensities of Sound

Noise levels are commonly measured on a *decibel* scale where zero represents the weakest sound that a young person with excellent hearing can pick up. Because there are sounds more than a thousand trillion times as intense—a large jet plane or rocket engine at takeoff, for example—the decibel scale is a logarithmic one. A sound ten times as great as another is 10 decibels louder; a sound 100 times as great as another is 20 decibels louder; and a sound 1000 times as great as another is 30 decibels louder. A change of one decibel is just detectable by the ear. The following table shows some approximate sound intensity levels on the decibel scale:

	DECIBELS
Threshold of hearing	0
Breathing	10
Rustling leaves	20
Library whispering	30
Talking in a quiet office or residence	40

	DECIBELS
Conversation in a classroom	50
Busy classroom discussion	60
Automobile engine running at a 3 meter distance	70
Some dishwashers and blenders	80
City traffic	90
Heavy trucks	100
Large jet overhead at 150 meters	120
Threshold of pain	120-140

Although younger students will have trouble mastering the concept of decibels, they can get a notion of the comparative intensities of sound. Several student helpers can print on individual strips of cardboard the sounds listed on the table, e.g., breathing, library whispering. The colored strips of heavy-weight paper available from Hammett or other supply houses and intended for use in pocket charts work well, for the strips come with lettering guidelines and are relatively stiff.

When the strips are ready, each strip is given to a student. The student holding the strip marked Threshold of Hearing and the one holding Threshold of Pain come to the front and stand about five meters apart. Students holding the other cards stand on a side of the room so that their cards are visible to seated students. Ask, Which sound do you think would be the hardest to hear and, therefore, the least intense? Who disagrees? When a majority agrees, the student holding the card moves into position on the sound intensity scale. Of course, disagreements will occur, and the scale developed may well differ from the one given above to some extent. For example, students will contend that not all library whispering is of equal intensity; some is louder than classroom conversation. Such disagreements will lead to the conclusion that only through accurate measurement of sound intensity can a true scale be developed.

Brrr-rrr-rrr!

After students have ordered the sounds according to intensity, the strips they hold can be mounted on a bulletin board captioned Our Sound Intensity Scale. Make additional small slips bearing other sounds such as water running from a tap, noontime noise in the cafeteria, a dog barking from three meters away, the grind of a pencil sharpener, ordinary footsteps, fire bell, school intercom, a scream, radio music played at normal volume, and an electric typewriter. Students can compare the intensity of these sounds to those on the sound intensity scale they have already developed and tack each new sound next to the one that appears to be approximately as loud. Again there will be different opinions, and the need for precise measuring instruments will be obvious. Also students may begin to realize that the same sounds produced under different acoustical conditions will reach the ear at different decibel intensities.

It may be possible to demonstrate a sound survey meter to your class. Persons who may use such meters in their work include health and environmental control agency officers, sound engineers, architects, physics and engineering professors, and acoustical material manufacturers and contractors. A demonstration may also give young people insight into environmentally related careers about which they may be unaware.

Perceptions of Sounds

Tuning forks can show that the human ear does not hear all frequencies (pitches) equally well even when the energy from each is the same. Strike a high frequency fork. Then hit a lower frequency fork with the same force. Ask students which sounds louder. They write their responses on slips of paper. Most students will select the higher frequency fork since a C_2 fork, which generates about 70 vibrations per second, will sound fainter to the human ear than a C_7 fork, which generates more than 2000 vibrations per second.

Participants can go on to generalize which frequencies may be most annoying to the human ear.

The first spoken sounds lost by the nerve-deaf are the fricative consonants: *f*, *s*, *th*, *ch*, and *sh*. This makes discrimination difficult among such words as flick, sick, thick, chick, and Shick. Greater hearing loss makes the explosive consonants difficult to distinguish: *p*, *b*, *t*, *k*, *g*, and *d*.

Investigators can make up word lists based on these groups of consonants. After practice, one student can recite the list in random order at a steady pace and volume while a noise source is gradually increased. Other students attempt to write down the words as they hear them. After several trials with differing background noises, a group of students can analyze the data using the following analysis guide to record data about the fricative consonants:

	lick	click	slick	quick	flick	Dick	thick	sick	Shick
A									
B									

WORDS

A Number of students not perceiving the word in a quiet room

B Number of students not perceiving word with background noise turned to high

Questions that should emerge from this activity include: Do background noises have the same effects on people's perceptions of these sounds? How well did students close to the noise do as compared to those at a distance? Did reflective surfaces such as walls have any effect on performance?

A student who has a relative who is hard of hearing can try the list on the person with and without use of a hearing aid, and with or without background noise. This can be a student report out of life instead of the encyclopedia.

The school nurse may prove a valuable resource in studying different perceptions of sounds. Ask the nurse to speak to the class and demonstrate the equipment used to test hearing.

Comparing Noises

Cassette tape recorders are very useful to record sounds under different conditions so that they may be played back later in rapid succession for ease in making comparisons. Be sure that the recording levels are kept constant so that fair comparisons are possible.

Record for about ten seconds the noises from point sources such as an idling motor vehicle, operating machinery, or banging of trash cans. Record at the same distance from each source. Now double the distance from the point source and record again. Are the noises just as loud, somewhat less, or do they appear to be less than half as loud as the first measurement? This comparison shows how distance from the noise source affects the sound picked up, but, of course, the results may also be influenced by such things as the amount of reflection and absorption present.

At a given distance record noise from a point source that can be moved, such as a transistor radio or the slap of a stick on a table. Then record the same sound in an identical fashion but in a different location. For example, investigators can compare noise levels in a large room vs. a small room, outdoors vs. indoors, in an empty room vs. the same room full of people, in rooms with and without carpeting and acoustical tile.

Some questions raised by these trials include: Does room size have an effect on how much noise is heard? Are sounds absorbed better by some building materials than others? Do people themselves absorb sound? Do some recording arrangements cause sounds to bounce (reflect) off surfaces and thus enlarge or lessen the annoyance of the sound? Is this the same for recording in all parts of the room? What difference does it make if an outside point source is in a large open area or near masonry building walls? In a room, what effects do open windows and doors have on the sounds received?

Generate a sound at the end of a passageway, alley, or other place where the sound will be channelized. Record the sound at a given distance and then again at twice that distance. Compare the sound levels received along the channel. Does the sound diminish? Some losses do occur as sound hits the walls, but, generally speaking, one can expect the loss of energy to be minimal. Investigators will probably have to go a considerable distance to record a lessening of channelized sound.

If the school is near a downtown, metropolitan area where the flat fronts of buildings form a wall-like façade, students might try the experiments successively in an alley, a narrow street, and a wide street. They can go on to consider the problems of trying to sleep when one lives in an apartment opening on an alley where sound is channelized.

Sound Barriers and Muffled Sounds

Are trees and shrubs effective as sound barriers? It will be difficult to make generalizations that answer this question. Growth and atmospheric conditions vary so much that even acousticians have difficulty with measurements. There is some agreement that a ring of trees will not block neighborhood noises from one's yard.

Students can test noise transmission through shrubs and trees and compare them to transmission across an open field. They might make a tape of a human voice and project it through the barriers to be tested to see under what circumstances words can be understood. They can try also a variety of low and high pitched sound sources. Reminder: when playing back a tape, always set the recorder at the same setting so that the sound source remains constant as it is heard through different sound barriers. A chart such as the following is a systematic way to collect data:

BARRIERS
Distance Between Source and Receiver ———— Meters

SOUNDS	THICK COVER OF TREES AND SHRUBS	THIN COVER OF TREES AND SHRUBS	GRASSY FIELD	PAVED DRIVEWAY
Human voice— tape recorder setting low				
Human voice— tape recorder setting high				
Barking dog— tape recorder setting high				
Rock music— tape recorder setting low				

In the blanks on the chart investigators can describe the sound (very loud, moderate, inaudible) as perceived across the barrier. Distance also can vary; in this case, recordings must be noted as distance between source and receiver changes from 5 to 15 to 30 to 50 meters. Use a meter stick to measure the distance; be sure that students record the distance being investigated.

Children shouting to one another across newly fallen, soft snow can notice the muffled quality of the sounds. If they examine a handful of the loose snow crystals, they can see the myriad of tiny air spaces that absorb the sound energy. Take an acoustical tile outside when your group investigates the muffling effects of snow so that a comparison can be made. Investigators can carry their field notebooks with them and in a few sentences write their observations, comparing snow air spaces and acoustical tile holes for spacing, shape, and size. Keeping field notes helps students to focus on the observation at hand, describe clearly, and write concisely.

Traffic and Neighborhood Noises

Hills, walls, and high earth berms or embankments along highways act as barriers to traffic noises and prevent direct sound propagation. Trees and shrubs help scatter the noises, but as students may have discovered in the previous activity, tree barriers are more effective with low frequency sounds than with high ones.

On a nonwindy day have students take positions in different noise barrier locations all the same distance from a highway. For example, some may be in a flat area, some with only dense shrubbery between themselves and the highway, others behind a hill or high berm. Listen to sound levels from different passing vehicles. If several tape recorders are available, different students can capture on tape the noise of the same vehicle under different sound transmission circumstances so that effectiveness of trees, shrubs, hill, and embankment can be compared as well as transmission from a flat area near the highway. Don't forget to set all the tape recorders at the same setting so that they record at the same volume level.

Repeat this type of activity on a windy day. Wind can carry noise into a former "sound shadow" area behind a hill or berm. Discussions can cover: How high must a hill or berm be to reduce sound to a comfortable or acceptable level? Are low and high frequency sounds lessened equally as well by the barriers?

Upper-grade students can conduct sound surveys in their neighborhoods to develop awareness of the noises that bombard them. After selecting a neighborhood site, each investigator writes down all the noises he or she hears during four predetermined ten-minute periods during a day and the length of time that each noise lasted. Students' records can include such sounds as the neighbor's meowing cat, the utility company's pneumatic drill, a bulldozer, an airplane overhead, a motorcycle, locusts, and the like. To facilitate the process, prepare dittoed recording guides for students to use:

SOUND SURVEY

Location of survey: _____

Date and time of survey: _____

Name of surveyor: _____

Noise	Source of noise	Length of time noise continued
_____	_____	_____
_____	_____	_____
_____	_____	_____

Students can repeat their surveys on several different days to find answers to questions such as: At what times of day do man-made noises tend to be greatest? Are there sounds that tend to occur primarily or only in the morning? at midday? at dusk? at night? Are there certain days of the week when man-made noises tend to be greatest? to be least?

Compiling sound surveys integrates easily with map study. Upper-grade students can sketch a large map of their community on a wall-sized chart and can plot each point in the community at which a sound survey was completed. The significant data gathered at each point can be hand-lettered onto small index cards and affixed to the map at the collection points.

Once the data have been mounted on the map, students can analyze the total data and draw conclusions. They can identify: 1) the major sources of noise in the community, 2) noisy areas in the community, 3) quiet zones, 4) noises that need to be regulated, and 5) necessary versus needless noise. Their conclusions can be compiled in a written report. The sound survey map as well as the written report can be sent to the city or town committee responsible for environmental preservation and future planning.

Noises in Homes and at School

Lower-grade students can undertake a more limited sound survey—a survey of their homes. They can construct charts on which they note the rooms and halls of their houses or apartments:

Address _____ Time Period _____		
Living room	Dining room	Kitchen
Bathroom	Front hall	Other

In each room block they list the noises produced *within* that room during the time period being investigated. Noises that probably will be included are refrigerator, TV, toilet, clock, talking, dog barking, running water, footsteps, and doors opening. Again, students should repeat their observations on several occasions, noting the time period under consideration.

After numbers of students have gathered data on several occasions, youngsters can work in groups of five or six to compile one large chart on which all noises heard in comparable rooms are listed. Then the children draw conclusions: What noises recur in living rooms? in halls? in bedrooms? What rooms tend to be the noisiest? the quietest?

Teachers who find that their students have difficulty making individual surveys can involve youngsters in team sound surveys of the school. Again the investigation can be integrated with social science map study. Participants can draw a map of the interior of their school. Then working in teams of three, each member selects a location to survey during four ten-minute intervals during the school day. When each team has compiled its data, it can be attached to the large-scale map of the school building. Students can follow up with conclusions developed through class discussion: What areas of the school tend to be the noisiest? the quietest? What noises in the school could we prevent? What noises in the school could we soften? Conclusions can be recorded on a large chart, and the accompanying sound survey map of the school can be sent to the principal or even to the board of education.

Noise Control Ordinances

Find out if your community has a noise control ordinance. If you can obtain a copy of the ordinance, young people can examine it with the following questions in mind: What kinds of noises does it include? vehicles? lawn mowers? machinery? barking dogs? aircraft? rail transport? horn blowing? sound trucks? How does the ordinance approach the problem of sound intensity? Is time of day a factor in determining what sounds are permitted? Does the ordinance appear adequate? overly restrictive? impractical? not clear? too limited?

If the ordinance is too technical or if a copy is not available, your students may be able to interview someone in authority who can explain it. Invite a member of the city or town planning board or a person in the community charged with responsibility for noise ordinances. Students can prepare for the session by listing the questions they want to ask; questions can include those noted above.

Upper-grade investigators can find out from the police department if it is the noise enforcement agency. Student interviewers can ask: What kinds of complaints are most common? What evidence must the enforcement agency obtain to present a legal case? How difficult has it been to get a noise pollution conviction? Again, investigators should prepare an interview guide before approaching the enforcement agency.

WRITING ABOUT NOISES

Describe the Sound

Ginn and Company (191 Spring St., Lexington, Mass. 02173) distributes a record titled *Sounds and Images* that can "kick off" an activity to build heightened awareness of environmental noise. *Sounds and Images* is a recording of simulated, nerve-jangling sounds, each of which can be interpreted in several ways. Students listen to several of the sounds. Then in three-person teams, they select one sound and write down words that describe it. Most words projected will suggest that the sound is unpleasant, reinforcing the notion that some sounds in the environment can irritate. The activity, in essence, is a brainstorming session that encourages students to play with words; it is a prewriting activity in which participants are projecting descriptive words that can become the vocabulary they can draw upon when they write descriptive paragraphs.

This activity can be transformed into a team game. Three-person teams listen to the same sound. As soon as the sound ends, teams list descriptive words. Only two minutes per sound is allowed for recording words. The team listing the largest number of relevant words wins the round.

Use in a similar manner several records distributed by Scholastic Magazines, Inc. (904 Sylvan Ave., Englewood Cliffs, N.J. 07632). *Documentary Sounds* supplies sounds of engines, machines, and so forth. *Sounds of Animals,* another 12-inch LP record, reproduces lion, elephant, hen, and other animal "talk." *Sounds of My City* presents the noises and rhythms associated with city living—subway trains rattling by, sirens blasting, and songs and talk of the people. Young children, in particular, will enjoy reacting to the recordings *Sounds of Animals* and *Sounds of My City.*

A class can listen to traffic noises and translate them into onomatopoeic words that literally express the sounds heard. Sound words discovered might include: *honk, boom, squeal, screech, clang,* or *chitty-chitty-bang-bang.*

This is a fun context in which to introduce the thesaurus. Pupils in upper-elementary grades can use Scott Foresman's *In Other Words* to find more words to express the sounds of traffic; they can go on to look through the sound section of *Roget's International Thesaurus* to uncover hundreds of clanking, jangling, clinking, and knelling-sounding words.

They can create their own portmanteau words *a la* Lewis Carroll—words formed by combining two words such as squeak and crunch into squnch, swish and buzz into swuzz, fizzle and wheeze into whizzle.

Sound Rhythms and Poetry

Writing poems in repetitive patterns is a natural next step to follow encounters with onomatopoeic sound words. Such poem patterns are built on the repetition of sound words as in:

> Rackity-tackity-tackity-rack
> Rackity-tackity-clackity-clack
> Rackity-tackity-tackity-rack
> Subway wheels strike the track

Give the youngsters a first line such as:

> Jingle, jingle, jingle, jang
> or
> Kerbang, kerbang, kerbang, crunch
> or
> Powie, bowie, bingo, bang

Begin by repeating the line rhythmically in unison three or four times. Youngsters in groups, as a total class, or individually can invent last lines, to add on after several repetitions of the line. As they gain sophistication, they can vary the beginning line as they repeat it a second or third time and then add their original last lines.

Because sound words themselves can have interesting sounds, acrostic poetry—often simply referred to as ABC poetry—can be rather effective.

Start the poetry writing with a brainstorming session. If the poem is to be about the speed and noise of trucks on a highway, begin by having class members call out *t* words that relate to trucks (terrific, trailer, tires, tremendous, tractor, transport, travel); then *r* words that relate to trucks (run, race, rock, roll, running board, road, roar, rush); then *u* words (under, up, ultra-large, ultra-loud, unloaded, unlucky, unnerve, unnavigable); then *c* words (cab, cover, carry, careen, clunk, clutch, crash, crush, crowd, carbon monoxide, cloud of fumes); and finally *k* words (king of the road, knock down, kilometer, kick off, keep). A recorder lists the words on the board. A researcher scans the dictionary for additional truck words.

The word TRUCK is then written vertically on the board, and each letter becomes the first letter in a line of poetry:

> Tractor trailers
> Roar
> Under
> Concrete overpasses,
> Kings of the road!

Brrr-rrr-rrr!

Students working in four-person groups select words from those the class has brainstormed, build them into noisy lines, and concoct five-line poems.

Other words that can be manipulated in a similar fashion include: car, bus, motorcycle, insects, rivers, waterfalls, lawn mowers, drills, motors, and bulldozers. Use the shorter words with youngsters in primary grades.

Students who have written repetitive sound poems can juxtapose words, noises, and poems on tape to create a "psychedelic" sound experience to share with other classes. Upper-grade young people can begin by collecting a potpourri of ear-shattering noises on tape; they retape the noises, interjecting noise words and noise poems shouted and whispered into the recorder. Their tape can make other students more aware of the cacophony of noises all around them.

Writing a Noise Control Ordinance

Upper-grade students may want to write a noise control ordinance for their own community, school, or classroom. Writing a classroom ordinance is a particularly good culminating activity for students who have participated in a number of noise pollution investigations.

Brainstorm as a class the question: What kinds of items should be included in our ordinance? Students will propose items related to the loudness of talk during independent study, work sessions, and class discussions. They may suggest the ordinance contain items about the loudness of the teacher's voice, about hall noise, about levels of audiovisual equipment, and about the loudness of the school's intercom and the use to which it should be put. Three students can serve as recorders during brainstorming, writing down the suggestions on paper. Their notes are given to a writing team, five young people responsible for drawing up a first draft. The writing team prepares multiple copies of the draft on a ditto. In a class meeting the draft is discussed and modified, then sent to an editing team, a committee that makes the suggested modifications and prepares a final draft to be ratified by the class. Once the ordinance has been approved by the class vote, students prepare multiple copies.

A meaningful speaking activity can develop as students take copies of their classroom ordinance to other classes and explain why each item was included. As students explain it to other students in the school, more awareness of environmental noise may develop.

CHAPTER 3

"THE AGE OF THE BIG THROWAWAY"
Activities with Solid Waste Disposal

A few miles outside Chicago, in Dupage County, there's a veritable mountain of garbage. Next winter people will be skiing on it. It is called Mount Trashmore, an affectionate term for a mountain of trash striding an area that was formerly called, with good reason, the Badlands.

Ski trails on a garbage dump? Sounds almost like a theme song for the Affluent Society, doesn't it. . . . Mount Trashmore is a dramatic example of perhaps the only nice thing one can say about pollution. Sometimes something can be done with it—not just about it.

> Keith Elliot, "Look What They're Doing with Trash," *Oilways*, no. 3, 1972.

Keith Elliot calls the times in which we live the "Age of the Big Throwaway" in an article on trash in *Oilways* magazine. He points out that in 1971 Americans threw away fifty billion empty cans, thirty billion bottles, four million tons of plastic packaging, and more than a million TV sets. How often do we use the expression "throw it away" without thinking where this "away" is? Is "away" the street? the garbage dump? a river? a marshland? the ocean? The truth is that we are burying ourselves in trash, filling in natural wetlands with garbage, and polluting harbors, rivers, the air in an attempt to rid ourselves of wastes. It seems obvious that we must generate less waste, reuse what we can, and dispose of unavoidable waste products in a manner that does not do irreparable damage to land and water.

Students can investigate some of the problems associated with litter and solid waste disposal and through their investigations, they may become more aware of the role they can play in alleviating the problems.

INVESTIGATING THE LITTER PROBLEM

Litter, Litter Everywhere!

Take a neighborhood walk to introduce young people to the problems of randomly discarded trash or litter. Armed with a clipboard, pad of paper, and pen, students can walk through their neighborhoods listing the specific items of litter they see. Their lists will include wrappers of all kinds, ice cream sticks, bottles, cans, cartons, and organic wastes from animals and plants. Older youngsters can convert their lists into tally sheets on which they check each recurrence of a particular item of litter and indicate the location at which the item was observed. Younger children can simply identify pieces of litter on the spot and follow up with a talk-time during which they consider where litter comes from, why people litter, and so forth.

After students have collected their firsthand data individually, in small groups, or as a total class activity, they can begin to process it by classifying the specific items of litter into categories. Possible systems for classifying that students can devise include:

☐ Size of litter: small, medium, large
☐ Kinds of litter: glass, metal, paper, plastics, organic wastes
☐ Prediction of how easily the waste will break down: slow, medium, or rapid decomposition

☐ Location at which the litter was found: open street area, wooded area, yard, vacant lot, and so on

This activity can help students comprehend the diversity of litter, the extent of it, and people's contributions to the littered environment.

Students can study litter more systematically by selecting a 50 x 50-meter plot near their school or home as an experimental site for intensive investigation. Armed with gloves, shovels, rakes, and receptacles in which to place collected trash, students can "attack" the chosen site and pick up the litter in the area. They record the following information about each item:

☐ Size: Between 0 and 10 centimeters in length
Between 10 and 50 centimeters in length
Larger than 50 centimeters in length

☐ Specific location where found: Use a gridlike map of the area to record locations

☐ Type: Paper, metal, plastic, wood, glass, organic plant, animal matter

Repeat the activity after two or more weeks have elapsed to determine the amount and kinds of litter that accumulate on a site during a specific time period. Students may wish to repeat the activity a third time to compare litter buildup during several equal time periods.

After students have collected data for a two-month period, they may decide to investigate the effect of PLEASE DO NOT LITTER signs on the amount and kinds of litter discarded on the experimental site. Students make signs, place them on the now litter-free experimental site, and then compare the litter buildup with the litter buildup during the same time period without signs.

Pictures Tell the Story

If the school owns a camera, students can photograph small

areas in which there is extensive littering. Keeping an accurate written record of the locations of picture sites, students can return to the areas on several successive occasions to take pictures so that they can compare the waste buildup and change over a period of time. Analyzing successive pictures taken at a location, young people can ask: What items of litter remain in the location? How do remaining items of litter differ from one picture to the next? What items of litter disappear? What hypothesis can we project to explain the disappearance? During what months of the year is litter buildup the greatest? the least? What seem to be the primary forces affecting litter buildup and removal at a location?

Students can compare the photographic evidence collected at several different locations and ask: What kinds of litter seem to disappear from all locations with the passage of time? Is litter buildup or removal greater at certain locations? What hypotheses can explain these differences? Are there different forces affecting litter buildup and removal in different areas?

Students can compose their pictorial evidence into a photographic essay. A description and an explanation of each photograph are written, revised, and printed on a card that is stapled onto a piece of black construction paper beneath the related picture. Each photo and description are mounted on a bulletin board with arrows connecting successive pictures taken at one location. If students prefer, they can bind the pages into an informational book.

This activity can be pursued as an independent project by one young person or by several students working cooperatively. Those who carry out the project can share their data and conclusions with their classmates by projecting their photographs with an opaque projector as they describe the changes they observed in locations after specific time periods had elapsed.

Photographs or drawings can also tell the story of a community's litter problems. Students shoot a roll of film or make sketches to produce a pictorial record of community sites that evidence litter buildup. Vacant lots, stream banks, shopping center parking lots, and abandoned buildings are just a few sites that can be recorded pictorially. A letter to the editor describing in words the specific problem sites uncovered by student investigators, accompanied by photographs or sketches, can be sent to a local newspaper for possible publication or to a community agency responsible for environmental protection.

If submission to a newspaper or community agency proves unfeasible, students can make posters with their pictures and devise a catchy caption to accompany each photo or sketch. A sketch of an abandoned and broken shopping cart, for instance, may be simply captioned "Why?" A picture of a vacant lot cluttered with discarded mattresses, bottles, and other refuse may be captioned "A New City Dump?"

Ban the Dog

TV commentator Jimmy Breslin has said, "Dogs were great in 1912. Now we have no room, and they are always underfoot. Curb your dog? Not near me. I want all dogs banned from the city."

Should dogs be banned from the city? Or should dog owners be required to clean up after their dogs as is mandated by law in Nutley, New Jersey? Or should there be no controls on dog littering in areas of high population density? The answers to these questions are subjective judgments, even though facts are cited by dog-banners and dog-lovers alike to support their differing points of view.

Divide your class into three teams: 1) For banning dogs completely from high density population areas; 2) For laws requiring dog owners to clean up after their pets; and 3) For freedom for dogs.

During a preliminary research period, students locate articles related to the topic, survey public opinion, and build an argument to support their team's point of view. Each team can list the specific points in its arguments on a large piece of paper. One student is chosen by the group to present the points to the total class during an informal hearing entitled Should the Dog Go? After the team presentations, all students can join the discussion of this highly controversial issue.

Johnny Horizon

A Johnny Horizon packet includes a plastic litter bag to be attached to a car or bicycle, a membership card in the Johnny Horizon Club, action and information booklets, and a Snoopy cartoon. The packet is obtainable in classroom quantities from Johnny Horizon, U.S. Department of the Interior, Washington, D. C. 20240.

Using the packet, students in grades four, five, and six can conduct a Johnny Horizon Cleanup Campaign. Directions for organizing committees, selecting a site, planning and cleanup, publicizing the project, and getting assistance in carrying out the project are outlined in the brochure *How to Conduct a Cleanup Campaign*. They can make posters using the facts given on the information sheet included in the packet, and make Snoopy cartoons, such as the accompanying example, that carry a message about littering.

Copyright United Feature Syndicate, Inc., 1971.

L Is for Litter

A creative writing and art activity can emerge from students' analyses of neighborhood litter. They can cooperatively write an ABC fact book on litter. Each youngster pulls a letter of the alphabet from a hat and writes and illustrates a page that begins with her or his letter. For example, the student who has drawn B may decide that B is for *bottles* and write a page about beer bottles thrown from car windows; illustrations can clarify the point being made. Very young children could simply find or draw a picture of one litter-related problem.

A letter such as Z has to be approached more imaginatively. Z may be about *zero litter*—the hoped-for goal; it may be for *zillions of kilos of waste*. A dictionary search can reveal other alternatives.

The individual pages can be bound together in a cover or held together with a ribbon or paper fastener. The book can be sent to the school library and checked out by other students.

FINDING OUT ABOUT REUSING AND RECYCLING

"Think Solid Waste Before You Buy!"

Consider for a moment the one-portion packages of breakfast cereal found on supermarket shelves. Each package has an inner layer of waxed paper and an outer cardboard covering. The eight to twelve individual packages are set in a cardboard tray and encased in a transparent paper wrapper that holds all the boxes together. Compared with the amount of paper required to package an equal weight of cereal sold in a bulk container, the amount of paper used in individual portion containers is excessive.

In its pamphlet *Waste Not, Want Not*, the U. S. Environmental Protection Agency suggests that buying more packaging material than needed costs money and makes more solid wastes. For example, it advises that, "Potato chips are cheaper when packaged in a bag," and that you shouldn't "pay for an additional box or can, unless you need the extra protection they provide." The Agency further advises:

> Check unit prices for the best buy. For example, cheese spread is packaged in a variety of ways. If it's in an aerosol can, you're getting mostly can. You're paying for convenience. Be sure it's worth the price to you. And be sure to recognize that, if your choice means more solid wastes, you should be willing to pay the price of proper disposal.

To encourage young people "to think solid waste" before they buy, assemble in class a collection of products that have been overpackaged—individual portion packages of cereals, aerosol cans of cheese, a box of potato chips with a foil lining, and the like. Students can describe and evaluate the packaging. If possible, display the same products packaged more conservatively and encourage students to compare costs.

Individuals or groups can investigate other instances of costly overpackaging. They can visit supermarkets to identify specific products, describe the way those products are packaged, and evaluate whether the products have been overpackaged and perhaps overpriced when compared with more conservatively packaged products of the same commodity. A chartlike guide can be a systematic means of recording data and of distinguishing between description and evaluation of product packaging:

A GUIDE FOR INVESTIGATING PACKAGING		
Product	Description of Product and Package	Evaluation of Packaging

This investigation need not be restricted to supermarket products. The packaging of McDonald's burgers, articles of clothing such as boys' shirts, toys, cosmetics, or nonprescription drugs can be investigated by student consumer researchers. Also researchers can investigate the percentage of container space actually occupied by the product. Is the box or container larger than necessary to hold and protect the product?

As a follow-up activity, students who encounter overpackaged products in everyday living can record descriptions and evaluations on index cards. The cards can be mounted in a column on a bulletin board, and eventually the information gathered can be compiled into "A Handbook for Environmentally Concerned Consumers," a mimeographed brochure that summarizes students' investigations.

In addition to making students more aware of the unnecessary packaging they buy, this activity requires students to distinguish between descriptive and evaluative thinking. Students first describe the attributes of a product's packaging and then make a judgment; that judgment is recognized as a personal evaluation of facts rather than fact.

Don't Throw It Away! Reuse It!

Although junkyards and junk collectors have existed for a long time, only in the last few years have citizens begun to participate in recycling efforts. When used glass is returned to glass kilns, when used metal products are returned to foundries, and when paper is sent back to pulp mills, the results are fourfold: 1) Natural resources are conserved, 2) Energy used in manufacture may be saved, 3) Garbage collection and dumping costs are reduced, and 4) The environment is less cluttered with waste products.

Another approach to limiting the amount of discarded waste and to conserving natural resources is to reuse items. Encourage students to imagine new and worthwhile uses of normally thrown-away items. To trigger positive suggestions for reusing materials, ask how the following items can be used and reused: large-sized grocery bags? small-sized grocery bags? plastic egg safeties? cardboard egg safeties? large plastic containers with handles that normally store salad oil, bleach, or antifreeze? jelly jars? baby food jars? What materials—if handled with care—can be used again for the same purpose for which they were originally intended, e.g., waxed paper, gift wrapping paper, plastic bags, aluminum foil wrap, rubber bands, or string.

This brainstorming activity can culminate in an ongoing bulletin board project. During independent study times, individuals or teams list on the bulletin board materials and the many reuses to which they can be put. A possible caption for the board is the familiar proverb "Waste Not, Want Not!"

Recycling Bottles

Glass container manufacturers in twenty-five states will accept color-sorted glass for reuse in their glass furnaces along with the raw materials—silica sand, sodium carbonate, and limestone. Metal caps and neck rings must be removed, but the glass need not be washed as long as it is reasonably clean. Reclamation centers will pay community groups about twenty dollars per ton for used glass.

Students will know if there are recycling centers in their neighborhoods. During hours when the centers are open and receiving glass from the community, students can ask and find out where the glass is sent and what the problems of collecting and transport are. Perhaps they can discover what percentage of the populace is participating in the salvaging effort.

In communities where there is no glass recycling program, students can find out the reason. They may discover that glass recycling is impractical in their community, because it is located too far from a glass factory, and transportation costs are prohibitive. An alternative is the returnable container.

Issues related to returnable bottles are good leads for student investigations that involve interviewing and letter writing as well as reading. Youngsters can try some of the following activities:

Ask the supermarket manager why "no deposit, no return" bottles are so popular with customers, storekeepers, and beverage distributors. Using a simple questionnaire, students can survey consumers' reactions by asking: Given a choice between a no deposit, no return bottle and a returnable bottle with a deposit, which do you generally choose? Why do you choose that bottle? If a number of students survey consumer opinion in this way, they can pool their findings and calculate the percentage of people in the sample that expressed a preference for each type of bottle. They can summarize the kinds of reasons people give to support their preferences.

Find out from a beverage bottler or distributor how many times a bottle can be used before it is beyond reuse. What problems are connected with collecting, washing, and reusing bottles? Students can compile their data in a flow chart that shows the steps in the reusing of glass and the problems encountered at each step.

Investigate the supply and availability of the raw ingredients that go into glass. Consider whether recycling glass actually saves valuable resources.

Debate the question, "Should the government mandate that *all* glass food containers—salad oil jars, applesauce jars, baby food jars—have a return deposit? Should all beverage bottles have a deposit to encourage return and discourage littering?"

Find out from the local department of public works—sometimes called the department of sanitation—or from the highway department what kinds of glass containers are discarded along roads and highways. Consider ways of discouraging people from such littering.

Select one product such as jelly, salad dressing, or peanut

butter, and visit the supermarket to study the shapes and sizes of the containers in which the product is packaged by different manufacturers. Record the data systematically by drawing in a field notebook the shapes of the containers and noting the weights or volumes packaged by different manufacturers. Formulate a judgment by asking, Are there too many shapes and sizes of glass containers? If shapes and sizes were standardized, would reuse be more practical?

Math-related projects for individual or group investigations include obtaining a more realistic conception of how much glass is used. Students keep a list for a week of the types and numbers of glass containers used in their homes, the types and numbers of containers reused, and the types and numbers discarded. A mimeographed recording guide similar to the following will systematize the investigation and make the ultimate pooling of results easier:

RECORDING GUIDE OF BOTTLE USAGE			
	DISCARDED	REUSED OR RECYCLED	TOTAL
Beverage containers (milk, soft drink, alcoholic, etc.)			
Wide-necked food jars (jelly, fruit, mayonnaise, pickles, etc.)			
Narrow-necked food bottles other than beverage			
Nonfood bottles and jars			
	Number of people in household _____		

The total number of bottles used in each household can be added together to give the total number of bottles used by all households in the survey. The total number of people in each household can similarly be added together to give the total number in the experimental group. By dividing the total number of bottles by the total number of people in the sample, students can determine the per capita use for one week.

The mathematical experts in the class can go on to find out the monetary value of the bottles used by an individual in a week. Begin with a round figure such as twenty dollars per ton for recycled glass. An investigator weighs several bottles of differing sizes and finds an average weight for a bottle. She or he then determines the value of one pound of glass (one cent) and estimates the value of the bottles used. Students who want to "go metric" may convert the units into the metric system.

Recycling Iron and Steel

The fall 1971 *Steel Facts* describes the mechanized refuse processing plant in Franklin, Ohio. The U.S. Environmental Protection Agency is helping this total recycling plant to develop and test means for "mining the urban ore." The system recovers paper fiber, ferrous metals, aluminum, glass, and agricultural compost.

Although steel cans constitute less than four percent of the incoming trash, iron and steel can be easily recovered, sold to a scrap processor, and end up in a nearby steel mill. One way that scrap iron is separated from other forms of solid waste is by a technique that depends on the unique magnetic property of iron and steel. Huge electromagnetic belts extract the iron from masses of trash.

Young people can demonstrate the process by making a simple electromagnet and testing its ability to pick up metal products normally discarded as trash, e.g., cans, bottle caps, foil, sheet metal, broken tools, and even discarded automobile parts.

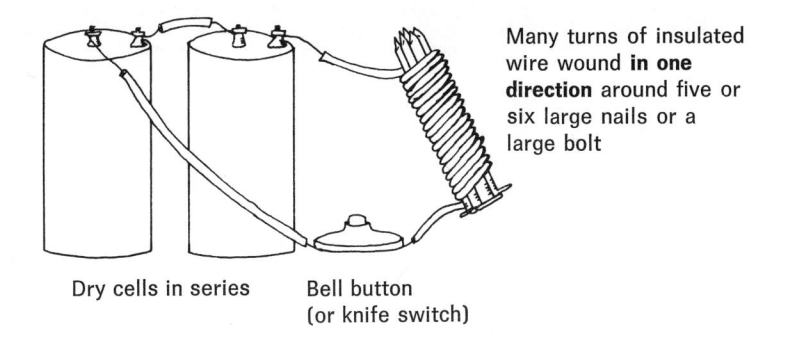

Many turns of insulated wire wound **in one direction** around five or six large nails or a large bolt

Dry cells in series

Bell button (or knife switch)

ELECTROMAGNETIC CIRCUIT

To make an electromagnet, wind a piece of insulated wire tightly around an iron rod or a handful of nails taped together. Make a circuit consisting of the wire wrapped around the iron core, two dry cell batteries connected in series, and a switch. The switch is essential because of heat buildup. Bring one end of the iron rod core near the material to be tested and briefly (to avoid heat buildup) turn on the switch. Is there an attraction? What kinds of materials can be separated according to electromagnetic properties?

For further information, students can write to: Public Relations Department of the American Iron and Steel Institute, (1000 16th St., N.W., Washington, D. C. 20036).

Much waste iron and steel find their way into automobile graveyards or junkyards. *Junkyard,* Nimbus Productions, a color, nine-ten-minute film, reveals the shapes and colors of junkyard wreckage. Viewing *Junkyard* is a visual experience for there is no narration—only scenes of wrecked cars, corrosion, and waste.

Youngsters involved in a study of solid waste management problems can react to the film by painting their imaginative interpretations of a world in which man is inundated by junk; fingerpaint is an effective medium for this type of creative expression. Similarly, students can make collages from materials generally thrown away to create scenes of an environment buried beneath heaps of trash. Don't forget the possibility of adding words that help communicate the message.

Recycling Aluminum Cans

In any total waste recycling plan, aluminum is an important component because of its high scrap value. Although by weight aluminum composes less than one percent of household waste, its resale value is over two hundred dollars per ton. Solid waste recycling plants hope to recover a large portion of their costs from the sale of the recycled aluminum.

The Aluminum Association (750 Third Ave., New York, N.Y. 10017) contends that recycling aluminum requires five percent of the energy required to manufacture new aluminum. It further contends that, compared to steel, on a product-for-product basis, aluminum costs about twice as much in energy to produce, yet aluminum, which is lighter, can be used over and over again for cans whereas steel deteriorates between recycled uses.

This issue—whether it is better to use aluminum or iron—presents upper-grade students with a complex problem that has several dimensions. Are the facts cited by the Aluminum Association about energy use in manufacturing verified by other sources? For what kinds of products are aluminum containers imperative? Is it better to use a product that has a high use-reuse potential, such as aluminum? Is it better, in these days of limited energy supplies, to use a product that requires a lesser amount of energy to manufacture? Considering questions such as these involve older students in judgmental thinking and the weighing of numerous pros and cons related to an issue.

Younger students may find it interesting to analyze the uses of aluminum in their homes. They can investigate the kinds of aluminum products used—pie plates, foil, and cans; the numbers of each product used during a given period of time, for instance, one week; the attempts being made in their own homes to reuse aluminum products. They can attempt a simple survey to find out how aluminum foil is used in households. They can ask the members of their families: For what purposes do you use aluminum foil? and estimate the number of times you generally use a piece of aluminum foil—once, twice, three times, or more?

Recycling Paper

Although paper comprises over half the trash collected by municipal sanitation workers, only about 20 percent of the paper annually consumed in the United States is presently recycled. Yet used newspapers, collected separately from other trash, can be deinked and made into new newsprint. Transportation and labor are the largest costs in recycling paper. Increasing the amount of reusable paper and cardboard will depend on the public's cooperation in making clean waste paper available.

To begin to comprehend the pervasiveness of paper in our society, students can look up *paper* in the yellow pages of the telephone directory. In addition to the listing of paper companies found under the general heading, they will find many other paper-related headings such as paper bags, paper boxes, paper-stock, and so forth. Items such as paper bags may be cross-referenced to other entries in the directory. For instance, in one telephone directory, the following entries were found:

Paper Bags
  See Bags—Paper
Paper Boxes
  See Boxes—Paper
Paper-Stock
  See Waste Paper

Following up the cross-references, students will find a host of related entries, for example, boxes—corrugated and fiber, waste reduction and disposal, and waste management.

Upper-elementary students can profit from a similar "trip through the yellow pages" as they search for names of firms they can contact to obtain firsthand information on paper uses and recycling. In so doing, they can learn the skills necessary for working with a directory that cross-references entries. If you are located in a small town, use the directory for a nearby city so that related entries will be more numerous and the process of tracking down references will be more complex.

Your class can also make an exhibit of items made from recycled paper. The exhibit can show cartons, posters, fiber cans, book covers, paper towels, plasterboard, and molded pulp items such as supermarket vegetable trays—all paperboard, the most important product of recycled paper. Perhaps you can borrow space from a local merchant or bank to bring the exhibit to the attention of the public. Students can add catchy slogans to gain public cooperation in local paper conservation or recycling efforts. Don't forget a credit sign for the class and school to accompany the exhibit!

Increasingly, paper and paper manufacturers are printing recycling symbols on their products:

RECYCLABLE PAPER

THIS PACKAGE MADE FROM 100% RECYCLED FIBERS

When putting together an exhibit, children may find it helpful to look for these symbols on paper products.

Is there a paper mill near the school that uses waste fibers? A class or committee visit might be a way of zeroing in on the conservation of paper. The Garden State Paper Company deinks old newspapers and makes new paper in mills in Garfield, New Jersey, Chicago, and San Francisco. Container Corporation of America has recycling mills in Los Angeles, Wabash, Detroit, Fort Wayne, Chicago, and Bakersfield, California. A building products company may be able to show your students waste paper being made into roofing and flooring papers, insulation board, or wallboard. Use the yellow pages to locate companies in your area.

After trying a number of paper-related activities, students will probably have developed a greater consciousness of the magnitude of the solid waste problem. They may show their increased awareness by generating less waste, taking more pride in preserving the quality of their environment, and cooperating in recycling efforts in their community. However, you may generate enthusiasm for community or school recycling action that has no means of being fulfilled. James Goetz of the American Paper Institute points out that separating paper from other solid wastes will not be practical in every community. A paper-stock dealer or mill must be available to guarantee a market for the paper collected. Also paper must be sorted and baled to have an economic value, and the shipping distance to a consuming mill must be economically practical. Young people should investigate these factors before deciding to undertake a recycling project.

Paper to Paper

When paper is recycled, it is placed in a large trough of water and fed under a heavy corrugated roller that beats and tears the wet paper into fibers of paper pulp. This machine is called a beater. The watery pulp fibers are then spread on a cloth or fine wire mesh, and most of the water is drawn off. The fibers lock together to form a new sheet of paper, which is then fed to drying rollers.

"Dorothy made this paper by hand from recycled fibers." All your students can experience the pleasure of printing a statement like this on paper they have made themselves. Experimenting with papermaking in the classroom provides a "hands-on" understanding of how discarded newsprint can be recycled into new paper. The basic steps in papermaking are:

1. Two small picture frames about 15 or 20 centimeters on a side are needed. Homemade frames of plain wood strips about a centimeter thick will serve but be sure the frames are not wobbly. Staple a piece of aluminum window screen over one frame. This is the mold. The other frame is placed directly on top of the screen side of the mold as a container to hold the pulp. Papermakers call it the deckle; it determines the size of the sheet being made.

2. Use an electric food blender as a paper beater. Fill it about three-quarters full of water and add small shreds torn from one newspaper page. Run the beater a half minute or so to make a smooth pulp. Pour the pulp into a plastic dishpan or other shallow container large enough so that the paper mold can be moved about in it. Make several more blender batches of pulp and pour into the dishpan; add about a cup of water for each batch to thin out the pulp. This will give a working quantity so that several sheets of paper can be made before one needs to make more pulp. Caution: *Carefully supervise any blender operation. Their knives are sharp.*

3. Stir the fiber batch well; fibers tend to settle to the bottom. Then grasp the mold and deckle together with deckle side up and slide them into the dishpan under the pulp. Gently rock the frames back and forth as you slowly lift them up through

the pulp. A smooth rocking action is necessary to spread the pulp evenly over the wire mesh and to allow the fibers to lock into one another. Hold the frames over the pan to drain for about thirty seconds.

4. Remove the deckle, and turn the mold over to transfer the pulp to a drying surface. (Don't worry—the pulp will not fall off the screen.) A smooth piece of felt or old worn blanket with little nap works well as a drying surface to insure that the new paper has a smooth texture. A square of heavy, old tablecloth or smooth curtain remnant can serve as alternatives. Several layers of newspaper could be used, but drying and separation are more difficult.

5. Gently peel off the mold and cover the pulp sheet with another layer of felt or blanket. Evenly applied pressure will hasten water absorption by the drying surfaces.

6. At this point an electric iron can be used to dry the paper, but it is not really necessary and wastes electric energy. Instead, when the new paper is still damp, peel off the top absorbing sheet and replace it with a plain piece of paper. On top of this, for a backing, lay a piece of glass or some kind of fairly stiff material such as heavy oaktag or a plastic placemat. Now lift off together the backing, the paper sheet, and the new paper, and turn them over so that the new paper is face up. It can now finish drying by itself or can be placed near a hot air vent or radiator to hasten drying.

Paper made by students out of old newspapers will be a light gray color when it dries, since it cannot be deinked first. The class can experiment making paper from other salvaged paper. Try used paper towels, manila paper, or construction paper. Adding a teaspoon of starch to each gallon of pulp will improve paper strength.

If you simply want to illustrate the fact that paper can be

pulped and the fibers reunited into a new sheet, dispense with the mold and deckle. Just lift out the pulp on a square of stiff screening and invert the screening on doubled paper toweling. Carefully press and blot with more toweling and leave to dry. A magnifying glass will show the interlaced fibers.

Students can write or type conservation or recycling messages on their sheets of recycled paper. The message can be prose or poetry. They can add the symbol that identifies recycled paper and sign the finished product.

In Philadelphia, the Franklin Institute operates a complete miniature paper factory where local or visiting students can see the entire papermaking process demonstrated.

An easy way to show reuse of paper products is to work with papier-mâché. Elementary school students can begin with used paper toweling and "recycle" it. They convert this waste material into puppet heads, maps, and models by first wetting paper strips with water and paste and then layering the strips onto the surface being covered.

Beware of the Hard Sell!

On a telecast of "The Advocates," J. P. Galbraith suggested that advertising serves two purposes—to inform and to persuade. Studying public relations brochures distributed by companies that manufacture products affecting the environment and by trade associations that represent those companies can help a young person perceive the techniques that such companies and associations use both to inform and persuade.

One technique is presenting only facts that support on the espoused point of view. Readers of such biased communications must be almost as aware of what facts are left out as what facts are included.

Public relations brochures distributed by plastics, paper, and glass companies, which support the use of their products as containers, are particularly suitable for a study of bias in advertising. The plastics people present the advantages of plastic containers over glass bottles, the glass people point up the disadvantages of plastic and the advantages of glass, and the paper people emphasize the advantages—rarely ever the faults—of their products. Only in cigarette advertising is a weakness or disadvantage of the product included, and that statement ("Warning: The Surgeon General Has Determined That Cigarette Smoking Is Dangerous to Your Health") is required by law.

Students can write letters to companies or associations related to container manufacture to ask for brochures describing the products and the companies' involvement in recycling efforts.

For glass write The Glass Container Manufacturers Institute, Inc. (330 Madison Ave., New York, N.Y. 10017). Names of specific glass manufacturers are supplied in a booklet titled *Glass Containers*; ask also for the complimentary *Litter Fact Book, The Solid Waste Fact Book,* and *Glass Container, the Ideal Ecological Package.*

For plastics write Mobile Oil Corporation (150 East 42nd St., New York, N.Y. 10017) and ask for *A Primer on Solid Waste.* Also Sinclair-Koppers Co. (Pittsburgh, Pa. 15219) and ask for *Plastics in Solid Waste—Facts You Should Know.*

For metal contact Manager Educational Services, The Aluminum Association (750 Third Ave., New York, N.Y. 10017). Ask specifically for the brochures *Litter, Solid Waste and Recycling* and *The Solid Waste Crisis,* which are sent in classroom quantities.

For paper write American Paper Institute (206 Madison Ave., New York, N.Y. 10016) and ask specifically for the brochure *The Paper Industry's Part in Protecting the Environment.*

For information about Oregon's bottle bill write Superintendent of Documents, Government Printing Office, (Washington, D. C. 20402) and ask for SW-109 *Oregon's Bottle Bill: The First Six Months;* there is a minimal charge.

When several brochures have been received, students can

compare the advantages and disadvantages of products as stated by a manufacturer and by a competitor. A reading guide may be helpful for students who have had limited experience with analytical reading:

PLASTIC VERSUS GLASS	
Advantages of Plastic Containers According to Plastic Manufacturers	Advantages of Plastic Containers According to Glass Manufacturers
Advantages of Glass Containers According to Plastic Manufacturers	Advantages of Glass Containers According to Glass Manufacturers
Disadvantages of Glass Containers According to Plastic Manufacturers	Disadvantages of Glass Containers According to Glass Manufacturers
Disadvantages of Plastic Containers According to Plastic Manufacturers	Disadvantages of Plastic Containers According to Glass Manufacturers

Students working on the project divide into four groups, each of which has the job of studying the brochures and completing one row of the guide. Each group can record its findings on a large chart for future sharing with the total class.

Young people who have investigated both sides of the plastics-glass container issue may wish to participate in a mock hearing modeled after the format of the TV program "The Advocates."

Several students can be witnesses for glass manufacturers with one student serving as the lawyer for the glass supporters. Other students can be witnesses for the plastics manufacturers and one can serve as the lawyer for the plastics supporters. Each lawyer in turn questions the witnesses, asking questions that will bring out both the pros and cons of the issue.

INVESTIGATING DISPOSAL AND DECOMPOSITION

What Shall We Do with Our Trash?

The unsightly mess of open dumps with their rats, flies, and offensive odors are still evident on the outskirts of many cities and towns. Dump fires to reduce the volume bring odors and smoke back to haunt those who contributed the garbage.

Sanitary land fills, as opposed to open burning dumps, are a partial answer to the question, "What shall we do with our trash?" At land fill sites, trash is covered over each day with soil or gravel and compacted by heavy machinery to reduce bulk. The covering keeps vermin out and prevents the entrance of oxygen, which could support combustion. Under these conditions, organic matter will slowly decay and return its mineral content to the earth; the material that breaks down is called biodegradable. Of course, metals not picked out by salvagers are lost; and since much of the contents of trash is not biodegradable, the land fill space is gradually filled. Moreover, the land fill itself may be destroying valuable marsh habitats, and pollutants may leak out into neighboring streams.

Students can investigate the final destination of the trash they are producing. In communities where there is garbage collection, students can interview the garbage man or the garbage contractor. Questions that can be asked include: Where does the garbage truck carry the waste? to a dump? to an incinerator? to a sanitary land fill? Is there any attempt made to salvage reusable materials such as paper or metals? If trash is being dumped, is there a problem of disposal space being consumed?

If students live in a rural area where families must dispose of their own trash, students can make a list of the things that can be done to reduce the amount that must be discarded or carried to the dump. They can visit the community dumping grounds to see if any attempts are being made to cover the garbage and to prevent fires.

When there is a sanitary land fill in the area, class reporters can determine if compaction is complete, or if air pockets are left; if the covering layer makes it impossible for rats to penetrate the trash; if garbage and trash pollutants are seeping from the land fill; and if some metals are salvaged, others buried.

A simple classroom experiment may expand students' concept of the term biodegradable. Bury small pieces of newspapers in moist soil placed in a series of convenient containers; margarine tubs serve well for this purpose. The tubs can be opened from time to time and the state of decay can be examined. Students can suggest different exposure conditions (a warm place, dry soil, rich soil) to see if the rate of breakdown changes. Younger students can include pieces of metal or plastic—nonbiodegradable materials—for comparison.

Decomposition—Nature's Slow Solution

Some forms of solid wastes decompose naturally, given favorable environmental conditions. Organic remains of a garden such as vegetable parts and tree leaves will be incorporated into soil after bacteria and other decomposition organisms have worked them over. How fast will newspaper, cardboard, or wax-coated paper products take to decompose? What conditions will be optimum for their decomposition?

A long-term comparative experiment can be carried out either at school or in the neighborhood to answer these questions. Students make litter bags of nylon or fiber glass. Fiber glass curtain scraps would do the job inexpensively. Flat bags made of two 20 x 20-centimeter scraps are sewn together with nylon thread or strands pulled from other fiber glass scraps.

Students place a number of layers of cardboard, paper, or milk cartons into the bags, sew them closed, and weigh each bag on a balance scale that can detect a difference of a few grams; remember that there are 28.3 grams to an ounce. This type of scale can be borrowed from the high school science department.

Place the sewn-shut bags in various environments to see how fast their contents will decompose. Ideally, a very damp, dark, warm place will encourage organism growth and more rapid decomposition. Bags can be pushed into decomposing leaf or needle layers of a woods, in closely piled trash, in a moist, grassy area, or even in a garbage dump, provided it will be undisturbed and can be located again. After three weeks to a month, collect the bags for study. They must first be dried out to eliminate any weight gain because of moisture. Then reweigh to find out how much loss there has been due to decomposition. To find the percentage lost, divide the loss of weight by the weight of the original dry litter bag and contents, and multiply by one hundred.

All students can make one or more bags to check. In each experimental location, several bags of different materials should be placed so that the effects of given conditions on different kinds of matter can be compared.

For ease in identifying bags as materials begin to decay, each should be labeled; a piece of aluminum foil on which a number has been scratched or a piece of plastic labeling strip on which a number has been imprinted are suitable for this purpose. Marking a number on a bag itself is less suitable because time and exposure to the elements will destroy the markings. Of course, the number affixed to a bag should be recorded in a notebook in which a description of the contents of the bag is written. Data collected each time a bag is analyzed are also recorded in the book.

Vary the experiment by placing bags containing the same kind of material in several different locations. Students can then compare the effects of different environments on the decomposition rate of a substance. Again careful records of both the environmental location and the material should be kept in students' notebooks.

This experiment can be continued with upper-grade students over an entire year; the bags are tested each month and placed back in the experimental location for further decomposition. Smaller bags would be more convenient to hide; but, of course, the weight losses would be proportionately less. This smaller loss would be acceptable if students are patient with the weighing process and have scales that can weigh to a fraction of a gram.

Don't expect vast differences in weight in a few weeks or months. If decomposition processes were fast, there wouldn't be the solid waste problem on the grand scale that exists today. The realization that decomposition proceeds slowly is a significant learning in itself and may affect behavior; students may be less likely to discard paper products carelessly.

Sewage—What Do We Do with It?

Many municipalities pass sewage through a secondary treatment that separates solids from liquids and converts the solids into a kind of sludge, which may be rendered harmless in the process. The municipalities must then dispose of the sludge. Incinerating, dumping in the ocean or a river, burying in a land fill, and converting to dry fertilizer are methods currently used.

Waste liquids have been used in farming. Over one hundred years ago the cities of Berlin and Paris piped municipal waste water to sewage farms where the waste was used to irrigate crops. Israel has been using the sewage of Tel Aviv to grow fruits and vegetables in the Negev Desert. An experiment in sewage irrigation is being operated by Penn State University, using the sewage from 20,000 students and an additional 10,000 townspeople. No evidence of contamination of well water by bacteria has been found near the Penn State fields because the soil acts as a filter. Plants use the nitrogen, phosphorus, and other nutrients from the sewage and thus help keep waterways clean. A plan under study by the U.S. Corps of Engineers would carry domestic and industrial wastes from Chicago to an irrigation site 25 miles south of the city. Muskegon, Michigan, is also gathering data at its waste water management project.

Sewage farming is only feasible on land that can absorb both sewage irrigation and natural precipitation; if the soil clogs or cakes, the process cannot work. The process is best for growing tree crops, cotton, stock feed, or vegetables and fruits that are always cooked or peeled before being eaten. Copper, zinc, mercury, lead, and boron from industrial sewage could poison plants or make crops unfit for human consumption.

Young people can investigate the means of sewage disposal utilized in their local community. If the community has a sewage disposal plant, the class can visit it or a student team can interview the engineer-in-charge or the plant operator. Does the sewage plant produce a wet or dry sludge? What does the plant do with the sludge? Is there any chance that the disposal of sludge is creating an environmental problem? Is the sludge being used for fertilizer or land fill?

If your school is located in a city, you may be able to contact a city official responsible for planning expansion or modification of the present sewage disposal system. The official may be able to visit your class to speak on plans to use sludge for fuel or fertilizer and on the way sewage is currently being managed.

Flow charts are ideal for recording the steps involved in local sewage management. Each step can be shown visually in a box connected to previous and successive steps by lines and arrows. By constructing a flow chart down a page, rather than across, students can also indicate alternate steps that could be taken to recycle sewage wastes; these steps can be depicted parallel to the actual steps taken in the local sewage management plant.

Flow charts are also an effective way to record the steps in sewage management used by cities of varying populations and in different geographical locations. Students can study an atlas or maps to identify specific communities that fall into the following categories:

Population: Below 10,000, between 10,000 and 100,000, between 100,000 and 500,000, between 500,000 and 1,000,000, and above 1,000,000.

Location: In the mountains, on the plains, in the desert, on the seacoast (ocean), on the seacoast (gulf), and on a major river.

A pair of students can write a letter to a community that falls into each of the categories; select at least one community in each category. The letter writers should ask for information about how the community manages its sewage wastes. The data contained in responses can be compiled in a series of flow charts and then analyzed collectively by class members as an exercise in projecting such generalizations as: How do larger communities manage their sewage? smaller communities? communities near bodies of water? communities in desert areas?

Students may discover that in some communities, especially ones with low populations, home sewage disposal is managed with septic fields and/or cesspools. The first student who gets a response with this information may undertake an individual project in which he or she finds out how a septic field and/or cesspool operates, sketches explanatory diagrams, and prepares an oral presentation for the class. In preparing the oral presentation, the student investigator may choose to tape her or his remarks rather than giving a "live" presentation to the total class. Other students can then go to the classroom listening center to view the charts and listen to the tape.

Students may be interested in pursuing innovative approaches to the problem of disposing of animal wastes. The Ceres Land Company in Colorado systematically collects the solid waste produced by cattle raised in pens, purifies it, and uses it as feed for the animals who produced it (see *New York Times* article, September 9, 1973, pp. 1, 48). The Ohio Feed Lot Company of South Charleston, Ohio, is making a profitable garden fertilizer from the purified manure of sixteen million head of beef cattle (see *Wall Street Journal* report, March 5, 1974). These operations or similar ones in your own area are ideal places for students living in the region to view modern agricultural practices.

Although students may at first be repelled by the idea of reusing sewage wastes as fertilizer, fuel, or even as animal food, consideration of such attempts may help young people understand that organic wastes do not necessarily have to be thrown away by a civilization that has the technology to put men on the moon.

Halting Pollution

As a culminating activity after young people have become highly conversant with the problems of solid waste disposal and litter, they can have fun writing parodies of popular songs. Using the music as well as some of the words, students can rewrite stanzas changing words to send a pollution message.

For instance, a stanza of "God Bless America" could be rewritten as:

> God bless America
> "Waste" high in trash,
> Trash behind her, before her,
> Trailing trash in her rivers and streams.
> From the mountains, to the prairies, to the oceans littered high.
> God bless America
> May we survive.

The popular Australian tune "Waltzing Matilda" can similarly be converted into "Halting Pollution." Others songs that are easily parodied are "Mine Eyes Have Seen the Glory," "On Top of Old Smoky," "America the Beautiful," "She'll Be Comin' Round the Mountain," and "Beautiful Ohio." Some nursery rhymes work equally well, e.g., "Little Miss Muffet," and "Little Jack Horner."

In writing parodies it is not necessary for lines to rhyme or even to maintain the same rhythm of the original rhyme or song. The message is the essential element.

CHAPTER 4

A THING OF BEAUTY OR DANGER
Activities with Water Pollution

... *water is a thing of beauty, gleaming in the dewdrop; singing in the summer rain; shining in the ice-gems till the leaves all seem to turn to living jewels; spreading a golden veil over the setting sun; or a white gauze around the midnight moon.*

> John Ballantine Gough, "A Glass of Water"
> (1817–1886)

In an article titled "Life on a Dying Lake" in the September 20, 1969, issue of *Saturday Review*, Peter Schrag described conditions in Lake Erie:

> Some thirteen million people live in the basin of this lake, 90 per cent of them on the American side, the rest in Ontario, the polluters and the polluted, perpetrators and victims, all of them dependent on a body of water that, according to the best evidence, is not yet dead but in danger. They drink its water, swim on its beaches, eat its fish, and sail from its harbors. At the same time, they, their cities, and their factories each day dump, leak, pipe, or drop into the lake several hundred million pounds of sewage, chemicals, oil, and detergents fouling beaches, killing wildlife, and imperiling the water itself. Sometimes you can smell and taste it as it comes out of the tap, sometimes you can see it on the beaches and often in the rivers —the Maumee, the Auglaize, the Ottowa—but most significantly, you fear, not what already exists, but what might—and could—happen if the process continues.

The thirteen million people living in the Lake Erie basin are not the only polluters of water supplies; the rest of us around the world are perpetrators and victims as well, all highly dependent on clean water, all with a personal stake in seeing that the pollution process does not continue.

To reverse the process requires changes in our ways of thinking and of acting. This is the purpose of the activities in this chapter. As young people investigate the water they drink and water environments and learn more about water pollution problems facing them, they may begin to see that what they do will ultimately determine whether water is a thing of beauty or a thing of danger.

INVESTIGATING DRINKING WATER

Particles in Water

Water carries particles of material in suspension and in solution. Particles held in suspension are relatively large and, therefore, are visible to the unaided eye or through a microscope. Suspended particles settle out of the water when left undisturbed and after a period of time accumulate at the base of the container. Particles can also be removed by filtering.

In contrast, particles held in solution are so small that they cannot be observed with either the unaided eye or the microscope. They are fine enough to pass through filter paper and do not settle out of the water holding them, even after extended periods of time have elapsed.

Because water contaminants are carried both in suspension and solution and because water purification must take into account both suspended and dissolved particles, an easy approach to a study of water contaminants is to begin with the processes of suspension and solution.

Students can analyze the rates at which suspended materials will settle out of the water. They measure equal volumes of different materials such as ordinary soil, coarse sand, fine sand, clay, sulfur, miscellaneous debris, powdered brick, or even iron filings. A tablespoon is a handy measure; one or two tablespoons of each material can be experimented with. Students fill a graduated cylinder with a certain amount of water; if a 1000 milliliter graduated cylinder is used, they fill it to the 600-800 milliliter mark. They add the solid matter to be tested, shake for several seconds, and then leave the cylinders undisturbed and perform two tests: 1) They hold each sample up to the light to observe any particles in suspension; and 2) They measure the rate of settling by recording the volume of material that collects at the bottom of the cylinder at selected intervals; a stop watch or a watch with a second hand can measure the time intervals. A table such as this is an aid for recording data in the second test.

MILLILITERS OF MATERIAL THAT SETTLE OUT

MINUTES	SOIL	COARSE SAND	FINE SAND	POWDERED BRICK	ETC.
1					
2					
5					
10					

If graduated cylinders are unavailable, students can make their own from olive jars; of course, they must first calibrate each jar by pouring in known volumes of material (5 milliliters, and so on) and marking the side with tape labels to indicate the specific volume levels.

A second investigative team can do the same experiment, but instead of leaving the cylinders of water and suspended matter

A Thing of Beauty or Danger 69

undisturbed, they gently shake the contents and measure the rate of settling at selected time intervals under conditions that simulate water moving in a brook. Students with a high degree of curiosity may vary the experiment by changing the amount of shaking given each sample; they agitate some samples rapidly, others less rapidly, and make comparisons. The team uses the same table as given in the preceding activity for recording data.

A third team can simultaneously study how suspended materials are removed by filtering. They fold filter paper and place it into a funnel. They add known volumes of matter to water, agitate well, and pour through the filter paper. The filter paper discs are then dried out; students measure the amount of dried matter caught on the filter paper and compare it to the amount of matter originally put into the water. Again they investigate a number of different materials such as soil, sand, debris, or sulfur.

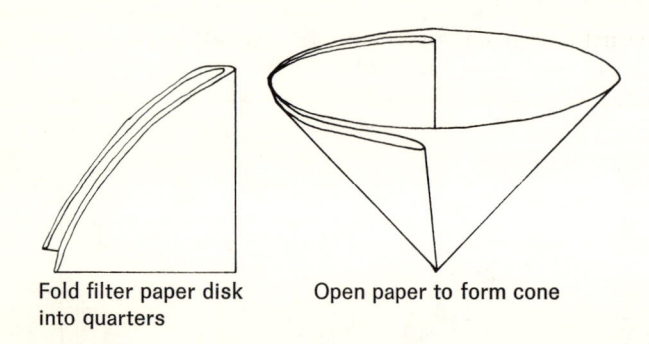

Fold filter paper disk into quarters

Open paper to form cone

A fourth team can perform a different but related task. Students put a pinch of cornstarch into water, agitate it, and test a small sample for the presence of starch. The test for starch is simple; starch in the presence of iodine solution will turn bluish. Students first discover this relationship by putting a drop of dilute iodine solution on a soda biscuit, piece of white bread, angel food cake, or potato. They then test their cornstarch-water combination and discover the same blue evidence of starch. Next they filter the cornstarch-water combination and test the liquid that passes through the filter paper—the filtrate—for starch. They will be able to categorize the starch-water combination as a suspension, because the filtrate gives a negative test for starch; no blue shows up since the starch settled and was filtered out of the suspension.

In a class reporting session teams share their data and conclusions. Questions to be considered at this point include: How do suspended materials get into rivers, lakes, and oceans? Under what conditions would a river or stream carry a lot of suspended materials? little suspended material?

If students live in an area close to a brook or river, they can conduct a firsthand study of suspended matter carried by water. They can collect samples of water and test it by: 1) Holding it up to the light to find out if they can see suspended particles; 2) Pouring each sample into a graduated cylinder and checking the rate of settling; 3) Pouring each sample through filter paper and visually studying the matter caught on the paper; then drying out the matter accumulated on the filter paper and weighing it to find the amount of material contained in that volume of water.

More Particles in Water

Students who have investigated the properties of suspensions can go on to apply their understandings and investigative skills to a study of solutions. Some possible studies are given below:

Put a measured amount of sugar or salt into a container of water and agitate. Allow the mixture to settle and then filter through filter paper. Test the filtrate by tasting to see if it retains the characteristic sugary or salty taste. Test foreign matter such as ink by adding a drop to water, agitating, filtering, and checking the filtrate visually for the blue color. *Conclusion* —when some particles are added to water, they dissolve; that is, the particles will not settle out and will pass through a filter paper. Two teams can experiment with salt, two with sugar, and two with ink. They all record their data for class sharing.

Weigh out four grams each of sugar, table salt, baking soda, and, if available, a colored salt such as copper sulfate. Place each in a beaker or flask. Add a little water at room temperature to each sample and shake well. Keep adding water, a small amount at a time, until all the solid is dissolved. Record the total amount of water added and compare the amount necessary to dissolve each material. *Conclusion*—materials differ in their degree of solubility.

Repeat the previous experiment using first ice water and then hot water to see if the temperature of the water affects the solubility of substances. Again add water, a little at a time, to each sample, shake well, and note the amount when no more water is needed to dissolve the test material. Measure and record the temperature of the water if a thermometer is available. *Conclusion*—the solubility of a material may be increased by an increase in temperature. This activity, combined with the previous one, can also be team investigations; two teams work with water at room temperature, two with cold water, and two with hot water; they record their data in tabular form, and a comprehensive table of results is compiled as students report their findings to the class.

MILLILITERS OF WATER REQUIRED TO MAKE A SOLUTION WITH FOUR GRAMS OF MATERIAL AT DIFFERENT TEMPERATURES			
MATERIALS	COLD WATER TEMP. = ___	ROOM TEMPERATURE WATER, TEMP. = ___	HOT WATER TEMP. = ___
Sugar			
Table salt			
Copper sulfate			
Baking soda			

Test tap water for dissolved materials by putting a small sample in an evaporating dish and heating until all the water evaporates. Check the dish when all the liquid has evaporated to see if there is a residue. If there is a residue, describe it. Does it have the reddish brown color characteristic of iron compounds or the whitish color characteristic of some salts? How do you think this material got into the water? *Conclusion*—water may contain dissolved mineral matter.

Follow up by mixing some soil into water, stirring or shaking, allowing to settle, and then filtering. Heat a small sample of the filtrate in an evaporating dish until all the liquid disappears. Now describe the residue. What does the residue represent? *Conclusion*—dissolved mineral matter in water may come from the soils that the water has been in contact with. Several students may choose this activity as an investigation during an independent study period.

Powder some limestone by grinding. Stir the powder into a sample of water, allow to settle, and filter. Heat some of the filtrate in an evaporating dish until all the liquid disappears. Describe the residue. Repeat the experiment with powdered sandstone, shale, or any rocks found in the area. *Conclusion*—dissolved mineral matter in water may come from rocks over which the water has traveled.

Add dyes to samples of water. Dyes can include food coloring, colored salts such as copper sulfate, potassium permanganate, or ammonium dichromate, and organic products such as beet juice, red cabbage juice, or cranberry juice. Filter each of the resulting liquids to see if solutions have been formed. In an evaporating dish or small saucepan, boil the filtrate down until all the liquid evaporates. Describe the residues. *Conclusion*—the color of water may come from dyes added to it or from natural organic materials.

Add salad oil to water to see if oil and water will form a solution. Shake well, hold the mixture up to the light, examine it visually, and describe it. Let the sample sit for a short time and examine it for possible settling action. Which material moves to the bottom of the container? Repeat the experiment using other varieties of oil such as light machine oil and motor oil; describe the results. Consider what happens when oil is spilled on water as happens when an oil tanker springs a leak. What is meant by an oil slick? *Conclusion*—oil and water do not form a solution; the oil comes to the top and floats on the water.

In each of the above activities, encourage students to formulate their own tentative conclusions. Don't tell them what is supposed to happen before they conduct the investigations.

Young people who have discovered some of the characteristics of solutions in their classroom laboratories may wish to investigate samples of water obtained from nearby streams, lakes, ponds, and/or seas for dissolved materials. As they collect their samples, they can observe the surface of the body of water for signs of floating oil. Collected samples can be analyzed by filtering and studying the filtrate visually to see if it retains any of the color of the original sample, as would be the case if the water has been contaminated by dyes; and by filtering, evaporating the filtrate, and describing the residue.

Samples obtained from various water sources or from different locations within the same source can be analyzed and compared in the same way. This would be a particularly interesting investigation if an industrial plant is dumping liquid or solid wastes into a community's river or lake. Samples of equal volume can be collected at a series of locations progressively downstream from the point of dumping. The samples are filtered, the filtrates examined visually and boiled to dryness, and the residues described. Older students may go on to weigh the residues to make more precise comparisons.

In this day of plastic bottles, students may wish to exchange water samples with students in other communities. Relatives or friends who live near other kinds of bodies of water can be sources of samples or students can send small samples of water collected in their geographical area to students in a distant community in return for samples mailed to them.

Life in a Drop of Water

Not only does water carry matter in suspension and solution, but it is home for scores of microscopic life forms not readily visible to the human eye. These life forms can be a source of diseases for people who drink the water. Study of a hay infusion is a fantastically interesting way to see the microscopic life that may exist in a drop of water and that may cause disease.

To make a hay infusion, simply put a handful or so of dried grass into a jar of water and store it in a warm place for several days. Timothy hay works particularly well, but dried grass from any field can serve too.

Students put a drop of the resulting infusion on a microscope slide and place a cover slip over it. If the infusion has stood for about a week, it should literally be loaded with microscopic organisms. Students can observe the organisms under a microscope and can sketch some of the life forms they see.

The life forms they will observe will depend on the source on which animals graze, dried grasses from a lawn, or dried weed stems from an abandoned field will produce infusions containing different forms of microscopic animal life. This life develops from eggs and cysts deposited on the dried grass; in the presence of the bacteria that grow in the organic, moist environment of a warm jar, the cysts open, and the resulting protozoa feast on the bacteria.

Making Water Fit to Drink

Settling, one of the processes young people encountered in their work with suspensions, is generally a first step in making water fit for human consumption. Water is pumped or run into large settling tanks where it is allowed to stand so that particles held in suspension will settle out. To make very fine particles of silt and clay settle, alum is added to the water. As the alum sinks to the bottom, particles of silt and clay held in suspension stick to the alum and are carried to the bottom too. This process is called coagulation.

Keep a large coffee can of silty soil in a laboratory center to which student teams can go during independent study times. They add about 600 or 700 milliliters of water to two 1000-milliliter graduated cylinders and a tablespoon or two of the silty soil and shake. To one cylinder they add a little alum—less than a teaspoonful will do—before shaking. While holding a white card behind the cylinders, they compare the clarity of the water samples after fifteen minutes.

Filtering is another step in water purification. Water is channeled from settling tanks into sand and gravel filters that remove most of the suspended matter. Students can make a simple working model of this process by placing a funnel in the mouth of a large bottle or flask. At the base of the funnel, they place a layer of clean pebbles. Above this, they place a layer of gravel, then a layer of coarse sand, and finally a layer of fine sand. Slightly muddy water is poured through the filter, while students observe the condition of the water that trickles through. Is this water pure enough to drink? What kinds of materials are still in the water?

Students who suggest that protozoan life may still be in the filtrate can rough-test their hypothesis by filtering liquid from a hay infusion through a pebble-gravel-sand filter, placing a drop of the filtrate on a microscopic slide, covering the drop with a cover slip, and observing their slides under the microscope.

To see how microscopic life can be rendered harmless to humans, students place a drop of dilute iodine solution or chlorine water to the cover slip of a slide containing protozoa. A

- Fine sand
- Coarse sand
- Gravel
- Pebbles
- Filtrate

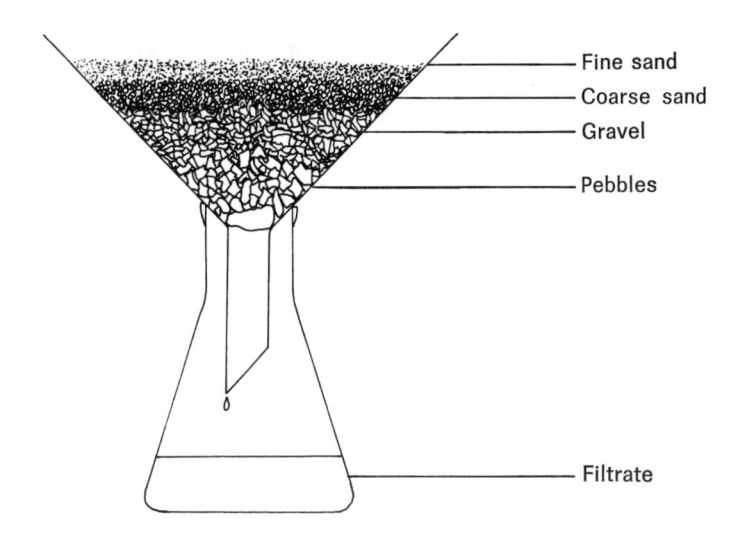

WORKING MODEL OF A WATER PURIFICATION FILTER

bleach containing chlorine can be used. As the solution spreads under the cover slip, students will see—through the microscope—the protozoa stop moving and will literally "see" why chlorine is added to drinking water in purification systems.

The hypothesis that filtering also does not remove dissolved mineral matter can similarly be validated by boiling some of the filtrate in an evaporating dish. A residue will be left in the dish after all the water has boiled away. Students may follow up by going home and studying the condition of the kettles used to boil water. If the water in the community is hard, students will probably find a residue in their kettles. Students living in homes with water softeners may investigate how softeners work and report to the class.

Two or three students may wish to set up a distillation device to demonstrate one of the best ways to produce water free of mineral material and animal life without adding chlorine.

A Thing of Beauty or Danger

A still can be made from a flask, one-hole stopper, glass tubing, collecting test tube, and water-cooling beaker as shown in the diagram.

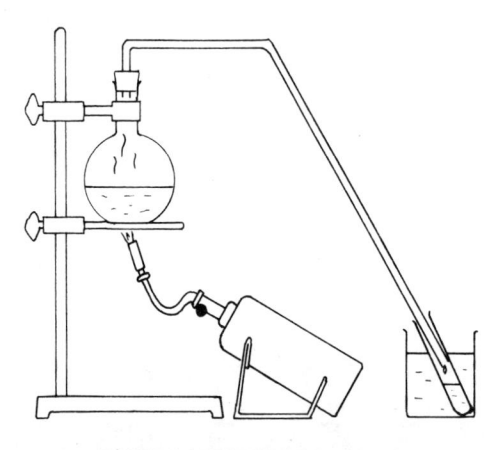

DISTILLATION APPARATUS

The sample to be distilled can be water that has been poured over powdered limestone, diluted ink, or liquid from a hay infusion. The distilled water can be viewed under a microscope to determine whether microscopic animals are still present, checked visually for the presence of the ink color, and evaporated to dryness to determine whether a mineral residue forms.

Students who have seen the magic of distillation may wonder why drinking water is not generally purified by distillation. This is an ideal time to consider with students the problem of the high energy required to complete processes such as distillation.

Once students have some understanding of the ways water is purified for human consumption, a few may want to undertake an investigative project in which they trace the water from their kitchen faucets back as far as they can go. To find out about the origins of the water they drink, student sleuths may have to contact the engineer at the local water purification plant for information or even visit the plant. Rather than compiling their data in a written report, students who select this project may present their data in the form of a working or nonworking model of the system or in the form of a large-scale diagram indicating steps in the process.

A useful resource for students undertaking this investigation is a nine-minute color film called *Water Purification*. Intended for youngsters in primary and middle elementary grades, it explains purification of water through simple demonstrations. *Water Purification* can be obtained on loan from Aims Instructional Media Services, Inc. (P. O. Box 1010, Hollywood, Calif. 90028).

INVESTIGATING WATER ENVIRONMENTS

Looking at Water in Nature

An observational study of the quality of water in a nearby brook, river, lake, and/or pond is a meaningful way to involve students with problems of water pollution. In conducting such a study, young investigators will learn some of the techniques used by environmental scientists in investigating water quality in an area, become aware of sources of pollution in their own community, and begin to realize that pollution affects them directly.

Boys and girls begin their study by observing the physical characteristics of the body of water in question. In cases of brooks and ponds, they walk along the banks; for larger bodies of water, they walk along a shore. They record:

☐ Amount and kind of floating material: twigs and other bulk organic debris, human litter, sewage wastes, detergent foam, scum, oil, and so forth

☐ Amount and kind of litter, debris, and wastes on the banks or shores

- ☐ Suspended matter
- ☐ Sewage, sludge, industrial dusts and fibers
- ☐ Evidences of water discoloration, possibly caused by dyes
- ☐ Gas bubbles
- ☐ Odors
- ☐ Visible algae and leafy plant growth in or on the water or growing from the banks or shores
- ☐ Locations at which materials are being discharged into the water and kinds and amounts of material being discharged
- ☐ Locations at which tributaries enter and distributaries go out
- ☐ Approximate width of a stream or diameter of a pond
- ☐ Condition of the bottom if visible: rocky, sandy, muddy, or sludgy

In noting what they see, students should aim for specificity rather than indicating that they "saw pollution."

Additional information about the body of water can be obtained from residents who live near the shore or along the banks. During observational walks, students can question people who are using the waters for recreational or commercial purposes: What specific evidences of pollution have you observed? What kinds of wastes do you discharge into the water? Have you ever been unable to use the water body because of excessive algal growth, unpleasant odors, litter, sludge, and so forth? Are these conditions worse during different seasons? different parts of the day?

Students can record information obtained through direct observation and talking with residents by making written notes. In *When You Are Alone/It Keeps You Capone* (Atheneum, 1973), Myra Cohn Livingston suggests a useful method for recording observations—a sheet with two columns, one headed *subjective*, the other *objective*. In the objective column go all the facts actually observed—in this case, the pieces of litter seen, oil slicks and dyes noticed, and excessive algal growth observed.

In the subjective column are recorded the feelings of the observers about what they see—"The smell made me feel as if my stomach were turned inside out," "Reminds me of a cesspool," or "Why do people do this?" The objective notes are compiled into a scientific report, and, as Ms. Livington suggests, the notes in the subjective column are the material from which poems are later made.

Students can also make a photographic record of the gross characteristics of the body of water. They can take pictures of debris floating on the surface, litter accumulated on the banks or shores, the wastes being discharged by an industrial plant, or activities such as swimming, fishing, or boating in the water. Hand sketches create a similar record. Then, too, students can carry a cassette tape recorder with them on their walks and record their observations on tape: "Two tires on the far bank, a discarded mattress spring lying on this bank, lots of organic debris on the surface, can't see the bottom. . . ." The tape is analyzed and data put into tabular form back in the classroom.

Still another method of recording data is mapping. Students sketch a rough outline of the body of water and plot observational data directly on the map at the appropriate points. They may devise a keyed code system for plotting data, e.g.:

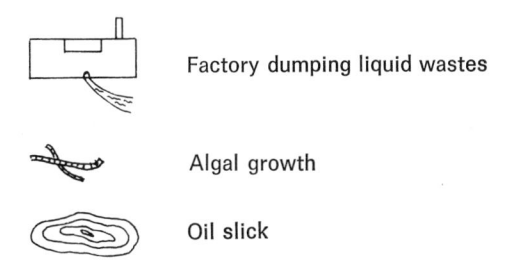

Factory dumping liquid wastes

Algal growth

Oil slick

To make their maps more accurate in scale, students can learn to pace off distances; they measure off a distance, perhaps

50 meters, in the schoolyard and pace that distance several times to determine the average number of paces they each take to cover 50 meters. When they have determined this, they pace off distances along the shore or bank of the body of water they are studying.

Or if students have a long tape measure, they can actually measure the length of the shore or bank. Sometimes a large-scale map can be obtained that shows the area being considered. Such a map can be used to make an even larger scale map of the water body. To do this, project the local map onto brown paper with an opaque projector and trace the outline. This enlargement may make the map more useful for observational recording.

How Deep Can We See?

Devised over one hundred years ago, the Secchi disk remains a valuable tool to the limnologist—the biologist who studies freshwater lakes and streams—and to the oceanographer—the scientist who studies the oceans. The flat, platelike Secchi disk with a connecting chain or cord is used to determine how transparent water is. It tells investigators how deep down they can see. Lowered into clear water, the disk is visible many meters down, while in water loaded with suspended matter, such as soil particles, organic debris, or plankton blooms, the disk disappears a few centimeters down.

It is relatively simple for even a beginner to use a Secchi disk. The correct technique is to lower it until it just disappears and record the depth, then to raise the disk until it reappears and record depth again. The final Secchi reading is the average of the down and up readings. Finding these depths is easy if the cord on which the disk is lowered has been marked with a felt pen at measured intervals. A ring is drawn around the cord at each half-meter interval and two rings at meter intervals.

Students in upper grades can use the Secchi disk for making

a number of comparisons: 1) Compare the transparency of the lake, pond, or river before and after a stormy period. Did the storm bring more suspended matter into the water? Did the extra flow stir up the sediments? 2) Compare changes in transparency during different seasons. Does a heavy growth of algae or plankton animals reduce visibility? 3) Compare the transparency of one lake with another. What factors account for similarities? differences? 4) Compare transparency in different locations on a lake or stream: near a roadway, by construction sites, at factory sites, at points disturbed by people and animals, at undisturbed sites.

Secchi disks are marketed under the names *visibility disk*, *limnology disk*, and *Secchi disk* by Macmillan Science (Chicago, Ill., and Boston, Mass.) Ward's Natural Science (Rochester, N.Y., and Monterey, Calif.), and by Carolina Biological (Burlington, N.C., and Gladstone, Oreg.). Disks cost about $15. When ordering, check if a chain for lowering the disk into the water is included in the price.

Do you have a student who would like to make a Secchi disk or a class parent who might volunteer to make one? Here is how to do it. Cut a 20-centimeter-diameter circle from a metal or Plexiglass sheet. Paint the circle white. Bore a hole through the center so that an eyebolt can be fastened through it for attaching the cord. Use an eyebolt long enough so that a number of heavy washers or nuts can be fastened underneath. The washers serve as weights to keep the disk horizontal as it is lowered into the water. Attach a cord to the eyebolt, and mark off half-meter intervals. By the way, some limnologists prefer a disk surface that is marked off into quarters with alternate quarters painted black so that when water contains white suspended matter, the disk is still usable.

You will find that students who have conducted elementary observational studies of a lake or stream will enjoy the experience of systematic analysis using a Secchi disk. Although the

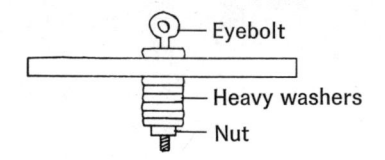

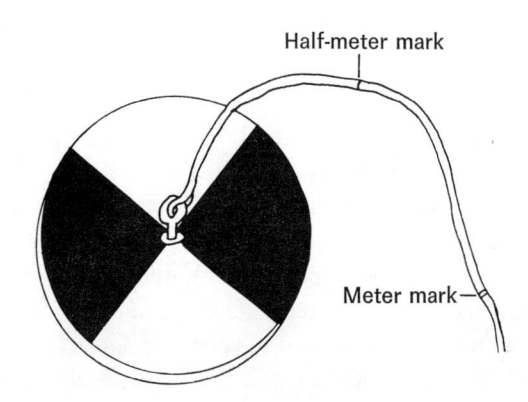

HOME-CONSTRUCTED SECCHI DISK

procedure is relatively uncomplicated, students will begin to feel that they are really scientists when they use one of the tools of environmental biologists.

How Fast the Water Travels

Students who find systematic study with a Secchi disk intriguing may enjoy making what limnologists call a "time-of-water-travel" determination. Time-of-travel tells a scientist how long it takes materials in a stream to move from one location to another. This is important as a scientist estimates the time it takes for waste materials to move downstream and the possible effects of such materials on organisms living downstream.

In *A Practical Guide to Water Quality Studies of Streams*

(U. S. Department of the Interior, Federal Water Pollution Control Administration, 1969) F. W. Kittrell suggests the orange as a handy float to use in determining time-of-travel of a stream. Oranges are round and are less likely to catch on protruding objects than an elongated float such as a twig. Oranges float partially under the water and, therefore, are less affected by wind than floats that ride higher in the water. They are colorful and visible. Also oranges decay; those not collected at the end of an experiment will not contaminate the environment as would a chemical dye.

Oranges are dumped into the stream being tested at a predetermined moment, and the time of their arrival at points downstream is recorded. Because orange floats may get caught in side grasses or in quiet eddy areas, investigators may have to throw sidetracked oranges back into the current. Kittrell points out that to toss a number of oranges into moving water and go way downstream to await their arrival may just not work. He cites a case where this was done, and not one orange from a crate dumped upstream arrived at the collection point downstream. All had been stopped somewhere along the way by eddies or vegetation.

However, a class will have little trouble overcoming this difficulty. Students can station themselves at 50-meter intervals along the stream, and when orange floats dropped upstream get caught in eddies or vegetation near their positions, they can free them and throw the oranges back into the current. Several students at key points—perhaps at 200-meter intervals—are armed with stop watches to record the times at which the orange floats pass their observational points. Of course, the time at which the oranges are dropped into the stream as well as the time when they pass the final collection point should also be recorded; students pace off the distance from the starting point to the final collection point so that they can estimate the time-of-travel down a known distance.

This investigation can be repeated under different conditions, e.g., immediately after a heavy rain, several days after rain, and during a dry spell. In this way, comparisons of the velocity of surface water can be made.

Again you will find that students will enjoy the systematic aspects of this activity. Although at first it may appear complicated, it is actually very simple to carry out, and is an event in which an entire class can participate.

Oxygen, Temperature, and Life

Young people who have conducted an observational study of a pond, lake, or stream and who have gained a little investigative sophistication working with a Secchi disk will be interested in a chemical analysis of water. A good beginning is to test for dissolved oxygen.

Because animals and many bacteria have a constant need for oxygen, the amount of oxygen found in water is an excellent indicator of the kinds of life possible and the biological activity in the water environment. Limnologists measure dissolved oxygen with an oxygen meter or by chemical titration. These methods are rather difficult and expensive for the nonprofessional; however, moderately priced test kits are now available so that a beginner with no chemistry training can measure the oxygen level of a brook, river, pond, or lake as part of an environmental study.

The Hach Chemical Company (Box 907, Ames, Iowa 50010) offers a dissolved oxygen kit for about $15. The kit is safe and easy to use since the chemicals for each oxygen test are supplied in premeasured packets. Each kit contains packets for one hundred tests; refills are available. Students can gain experience by testing the dissolved oxygen levels in the classroom aquarium, in tap water, and in bottled water. Directions are right on the kit.

To learn what is a high or a low oxygen reading, students can conduct the following experiment: *1)* Put at least a liter of water in a container; *2)* Oxygenate it with an air pump or by pouring the water back and forth from one container to another; *3)* Boil an equal amount of water for about ten minutes to drive out the dissolved oxygen; *4)* Cool the boiled water to the same temperature as the oxygenated water; *5)* Use the oxygen test kit to measure the oxygen level in each vessel. Ten parts of oxygen per one million parts of water is considered high; one or two parts of oxygen per one million parts of water is low.

To see the effects of oxygen level on an organism, place a goldfish in the oxygenated water and one in the cooled boiled water and observe. The fish in the cooled boiled water will probably come to the top gasping for oxygen. The fish in the oxygenated water should not behave this way. Now students can hypothesize what the fish behavior will be if the gasping fish is placed in the oxygenated water and the other fish placed in the deoxygenated water. Students make the exchange to verify their hypotheses. Because you don't want the fish to suffer, it would be best to conduct this experiment as a total class experience. That way a gasping fish can be placed in a more healthful environment as quickly as possible.

Now encourage students to transfer their understanding of how to measure the dissolved oxygen level of water to a study of bodies of water in their community.

The first job is to collect samples of water. To collect from the bottom of a pond or stream or from some level below the surface, an empty container must be lowered to the desired depth, filled, and raised to the surface. A serviceable collecting container, called a Meyer bottle, can be made from a plastic gallon jug in which products such as bleach, ammonia, cider, or milk are packaged. Of course, a glass container can be used, but the danger of breakage makes plastic a wiser choice.

To make a Meyer bottle, fit an eyebolt through a cork or one-hole rubber stopper. Fasten a cord to the eyebolt on the stopper and then to the bottle handle with a little slack. Plastic-covered clothesline makes a strong cord. Tie a brick or piece of scrap iron to the handle as a weight, so that the bottle filled with air can sink to the bottom. Or, as a refinement, fasten an old iron gear or other piece of metal permanently to the bottom of the bottle. This is less clumsy and keeps the bottle neck upright and out of the bottom mud or silt.

To use the Meyer bottle, stopper it tightly, lower it to the desired depth, and with a quick jerk, pull the stopper so the bottle can fill. Wait a few minutes. Then pull up the bottle by the cord that is fastened securely to its handle. Take a sample from the Meyer bottle to test for dissolved oxygen.

There is one problem in collecting water samples this way; oxygen from the air in the bottle may dissolve in the water being collected. Thus an oxygen test can register a higher level than is actually present in the pond or stream water. Professionals overcome this limitation by using a more elaborate device called a Kemmerer sampler, which prevents air contact.

Once students have constructed their Meyer bottles, they can use them to conduct a number of different tests. A thermometer placed in the bottle before it is lowered into the water can register the temperature of the water at a desired depth. Of course, the investigator must allow time for the thermometer in the Meyer bottle to register the temperature of the surrounding water before pulling it up. If investigators mark off half-meter intervals on the cords attached to the bottles, they can easily determine the depth at which the temperature is measured.

In addition, students can use their bottles to collect samples of algae and/or minute animal forms living at the sampled depth. Since most of the life forms will be almost microscopic, they will have to pour some of the sample into a small glass jar and hold it to the light for viewing. Suggest to students that they also use a magnifying glass to study the samples.

Using a combination of the tests described above, more

sophisticated student investigators can study water circulation in a lake. During fall and spring in temperate climates there is good circulation of water from top to bottom in many lakes so that both dissolved oxygen levels and temperatures are about the same at different depths. In contrast, during the summer many lakes, particularly deep ones, have a warm surface layer overlying a colder, deeper layer. Near the bottom there may be considerably less oxygen as bacteria and animals of the ooze and mud use it in their metabolism. Students can look for these seasonal patterns by checking samples of water from different depths for temperature and oxygen. Students who go to camp in the summer may carry out this investigation and take thermometers and their homemade Meyer bottles to camp with them.

Life in a Healthy Stream

Healthy streams have enough dissolved oxygen to support a variety of small organisms. Many insects lay their eggs in water where they hatch into larvae that spend a season or two developing into adults. Mayflies, stoneflies, dragonflies, damselflies, caddisflies, and others in the larval form provide a basic food supply for fish in streams that have an adequate oxygen level.

In contrast, the oxygen available in a polluted stream is usually very low and cannot support many of these larval forms. The variety of larvae present is one indicator of how healthy or polluted a stream may be. Here are some techniques for studying life forms in healthy streams:

Wade into a shallow stream that has a stony or pebbly bottom and is only twenty to thirty centimeters deep. An investigator can observe by standing quietly so that the bottom is undisturbed and the water retains its clarity. Larvae may appear on the stones or pebbles. Many will be only a centimeter or two long, some even smaller.

Pick up a rock or submerged stick and hold it to the light. As water drains away, "things" may begin to wiggle. Mayfly and stonefly larvae are commonly found this way. If little tubes of dark webby material are on the rock, a gentle squeezing of the webby stuff may force a caddisfly larva from its housing. Some caddisfly larvae build webby houses; others build tiny conelike houses from sand grains or bits of plant debris.

Catch larvae and other small animals on a wire screening. A collecting screen can be made from a rectangular piece of aluminum window screen about 30 x 60 centimeters. It will be easier to manipulate if the short sides are stapled to smooth wooden support rods or dowels that extend to form handles. One student faces upstream and holds the screen in a low slant *against the bottom*. A partner agitates the sand, pebbles, stones, and so forth upstream. Dislodged debris and organisms are swept onto the screen by the current. The screen is quickly lifted out and immersed in a catching basin such as a plastic dishpan or white enameled tray.

The choice of an appropriate dip net or sieve net is important. Delicate mesh nets of the kind used to catch butterflies will be torn by sharp stones or snagged on sticks or other debris of streams and ponds. A mesh net to catch bottom organisms should have one flat side on the rim and a canvas collar to protect the delicate bag. Such a net can be moved slowly along a pebble or sand bottom and gently agitated to loosen organisms. Or one student can hold it against the bottom while a partner turns over stones upstream. If you want to order a collecting net, ask for a D-frame aquatic net from Macmillan Science or Ward's; it should cost about $15.

For rougher work, a wire sieve net is more useful. It is essentially a sturdy rectangular wire basket supported on a pole handle. It can be swept through plant debris, mud, or pebbles in the water, or it can be pulled upside down, like a garden rake, from the shore. An assortment of nets is pictured in the catalogs of such science supply houses as Macmillan, Carolina, and Ward's Natural Science. A large, coarse kitchen strainer firmly attached to a pole makes a workable substitute.

Good nets should be washed in fresh water and carefully air dried after use; remind students that clean-up procedures are a necessary part of field and laboratory work.

Look for larger life forms in deeper pools carved by rushing water. These pools sometimes provide refuges for fish, crayfish, and frogs. A quick, careful move with a cloth net may capture one of them. Remind students that the purpose is not to see how many organisms they can take from a habitat, but rather to find out what kinds of organisms frequent which habitats and to get some idea of their abundance and variety as measures of the health of the environment.

Scrape green or golden brown film from the rocks. Scrapings can be put into a white enameled pan and examined with a magnifying glass. Patches of green or masses of green threads will be algae. Through their photosynthetic activity, they provide oxygen to the water. To see the individual cells of the green algal filaments or the many diatoms common in the brownish film found on rocks, students will need a microscope that clearly magnifies at least one hundred times.

A Guide to the Study of Freshwater Biology by James and Paul Needham (Holden-Day, 1962) is a useful paperback to aid identification of algae and diatoms as well as the other forms of life discovered in the above activities. The book has a rich variety of diagrams and descriptions of life forms; it includes suggestions on collecting and analyzing aquatic organisms and their environments.

Laurence Pringle's *This Is a River* (Macmillan, 1972) is also a good reference for investigators in middle and upper elementary grades. It gives descriptions, photographs, and sketches of some of the plants, insects, and fish that live in rivers and streams and tells where flowing water comes from and where it goes.

In a Running Brook by Winifred and Cecil Lubell (Rand McNally, 1969) is a third good reference book for youngsters. Its detailed drawings by Ms. Lubell will aid identification, and the highly readable text contains fascinating information and suggests ways of observing the tiny creatures who inhabit fresh brooks and streams.

Although students should not remove numbers of organisms from their natural habitat, upper-grade investigators may wish to study movements, feeding habits, and body features of selected animals in the classroom. More can be learned if animals can be examined quietly under good light; however, it is not easy to keep all forms alive out of their natural habitat. Animals accustomed to rushing waters may not get sufficient oxygen when placed in a bowl or jar.

Some precautions to take in bringing back live specimens are the following. Avoid mixing different animals in the same container. Predators such as the dragonfly larvae may eat others such as the mayfly. Avoid mixing delicate organisms with rock samples; delicate organisms may be crushed if stones are brought back in the same container. If closed containers are used, remove the tops as soon as possible, and keep jars refrigerated or cool overnight if the organisms will not be examined until the next day. This will lower cell activity and reduce oxygen demand. Some insect larvae such as *Simulium*, the blackfly, will probably die despite this precaution.

Specimens can be preserved in five to ten percent formaldehyde or in seventy percent alcohol, but it is better to examine them while they are fresh since preservatives tend to make them soft and fragile.

Life in Polluted Streams

Students who have investigated the physical characteristics of a stream may have found evidence of organic pollution. Such characteristics as foul odors and organic sediments on the bottom or in suspension are indicators of large bacteria populations. As bacteria utilize the organic matter, they can rapidly deplete dissolved oxygen in the water and cause changes in the forms of life in the stream. Students can find evidence of change by examining rocks from the stream bottom to see if they have stonefly, mayfly, or caddisfly larvae. These insects are most sensitive to organic pollution and may not be found in contaminated streams. They can examine rocks in very shallow but fast-moving water to search for clusters of bulbous-looking larvae about 7 millimeters long fastened to the upper surface of rocks. These are blackfly larvae. The blackfly as well as horsefly larvae are less sensitive to organic pollution and may be found in larger quantities in polluted water.

Students might also scoop sludge or organic ooze coating the bottom to search for sludgeworms and bloodworms commonly found below a sewage outfall. They can nail a can on the end of a pole to reach the organic matter, dump the matter into a white enameled pan or plastic basin, and examine the matter with forceps or short sticks. Investigators may find red tubifex worms that have a snakelike movement and are highly tolerant to organic pollution. To identify their specimens, students can compare their catch to tubifex worms purchased from an aquarium supply store where they are sold as live food for tropical fish. Bacteria and sewage worms thrive where there is high organic matter and low oxygen and may be found in very large numbers since these conditions are ideal for them.

Reminder: Students should wash their hands thoroughly after examining sludge with forceps; they should not touch this matter directly.

Life in Lakes and Ponds

While there is no clear-cut distinction between lakes and ponds, lakes are large bodies of water having shores without vegetation because of wave action. Ponds, in contrast, are small, quiet water bodies with vegetation growing down to the edges and with rooted vegetation growing in the shallower parts or across them. Using these general characteristics, students investigating a body of water can determine whether they are dealing with a lake or pond.

The actual investigation can start with a walk along the strandline on the shore of a lake where the wind and wave action tend to deposit bits of debris. Students record signs of a normal or a polluted habitat using the following table.

INDICATORS OF NORMAL AND POLLUTED HABITATS

A NORMAL HABITAT	A POLLUTED HABITAT
Bits of plant matter and an occasional tangle of algal threads	Massive clumps of drying and decaying algae
An occasional dead fish	Large numbers of dead fish indicating a fish kill from a sudden polluting occurrence
Tiny amounts of debris of a variety of kinds	Heaps of garbage, paper wads, styrofoam chunks, and/or other litter that has floated
Scattered spots of foam	Heavy scum or oily and tarry residues left by wave action

Students can next look for signs of eutrophication. Lakes and ponds reach a state of eutrophication when they evidence heavy algal growth. The burgeoning algal blooms are blown toward shore where they decay causing disagreeable odors and attracting insects. The process of decay depletes the oxygen dissolved in the water and needed by other organisms that are a source of food for some fish. Heavy algal growth, odors, and insects are signs of eutrophication.

Where students do find evidence of eutrophication, they can investigate the origin of the condition. The ultimate cause, of course, is the nutrient richness of the water environment resulting from high levels of phosphates and nitrates. Students can look for ways these phosphates and nitrates enter the water. Is a sewage disposal plant discharging nutrients derived from organic wastes and detergents after incomplete treatment? Are septic tanks draining into the water? Is raw sewage being dumped from toilets on boats? Are any industries dumping wastes including phosphates and nitrates into the water?

The investigation can move to an examination of a sample of bottom sediment. For sampling soft bottoms, freshwater biologists use an Ekman dredge. This instrument has jaws that shut and grab mud when it hits the bottom. Students can improvise by dragging a weighted metal bucket or can over the bottom with a strong nylon rope or clothesline.

The collected mud or ooze is poured slowly through a fine wire screen to separate out any organisms that live in or on the mud. Snails, freshwater clams, and wiggling phantom midges (*Chaoborus*) may be found. Phantom midges are hard to see but interesting because, although they are only about 8 millimeters long, their transparent bodies look glassy against a black mud background. They are the larvae of a fly and are found in organic-rich lake and pond mud. Students should not conclude that bottom ooze necessarily represents pollution. Lakes and ponds normally accumulate organic mud over the years from the remains of plants and animals.

Students may also be interested in capturing minute swimming organisms. Particularly during late spring or early fall many lakes and ponds have a large population of organisms that are barely visible to the naked eye. These animals and plants may be carried by the water current despite their own ability to move. For this reason they are called plankton, which means drifter.

The forms of animal or zooplankton that students will find are primarily small crustaceans including water fleas and their relatives such as the one-eyed cyclops. A magnifying glass or a low power microscope is good for examining specimens found. The previously mentioned *A Guide to the Study of Freshwater Biology* is helpful in identifying specimens.

To capture plankton for examination, one student holds a plankton net with its opening horizontal while another pours ten or fifteen buckets of pond or lake water into the net. The bottle end of the net in which the organisms have been captured is removed or the net is turned inside out into a bucket of water. Or one student can work the net alone by walking up and down the shore dragging the net through the water. This is facilitated if the towing bridle of the net is attached to the end of a pole so that the net mouth faces the water as it is pulled. Biologists often tow their plankton nets through the water behind a boat.

Professional plankton nets can be obtained from Macmillan Science or Carolina Biological for about $27; Macmillan also sells a smaller net for under $15. Students can improvise a net by cutting the toe from a woman's discarded nylon stocking, fastening a small jar or vial to the toe end, and fastening a hoop of wire to the larger, upper end. When such an improvised net is used, the water passes through the stocking net, and the organisms are flushed into the collecting jar or are held against the inside surface of the net.

In addition to studying the minute living organisms, students may look for larger animal forms that make their homes in a pond or lake—frogs, salamanders, crayfish, turtles, beavers, or muskrats.

A fine little book for students to take along when they look for organisms is *The Observer's Book of Pond Life* by John Clegg (Warne & Co., 1956). It fits into a pocket and is filled with photographs and diagrams of water plants and animals that student investigators can use to identify their own specimens. Younger students may take *The Ladybird Book of Pond*

Life by Nancy Scott (Wills & Hepworth, 1966) with them when they go out to a pond for observations. Color drawings and explanatory statements are a guide to things to look for.

Another fantastic reference is an article called "Teeming Life of a Pond" by William Amos in the August 1970 issue of *National Geographic*. There students will find magnificent color plates of such organisms as water striders, dragonfly and damselfly nymphs, bloodworms, tadpoles, diatoms, and water fleas. A three-page foldout portrays over fifty life forms found in a pond. Amos's book *The Life of the Pond* (McGraw-Hill, 1967) describes creatures of North American ponds and is a superb reference as well.

Chemicals in Our Water

Phosphates and nitrates are the key nutrients necessary for the growth and reproduction of algae and the small water organisms that feed on the algae, but when the supply is enriched by domestic or other wastes, an enormous population expansion may result. The algae blooms become so numerous that they clog the water; when they die, the water is fouled when bacteria act upon their remains and consume the oxygen in the water.

Students can test for the presence of these nutrients by using commercial kits. Because they are usually designed to test for amounts of nitrates and phosphates found in natural waters, classroom investigators will have to dilute solutions of the compounds to simulate lake or stream conditions and to make tests.

This is easily done. Students fill a liter container with tap water and add a gram of a nitrate or phosphate compound. They may use a chemical such as sodium phosphate as a source of phosphate, or they may locate a phosphate-containing detergent or garden fertilizer by checking labels. Actually the latter sources may be best because detergents and fertilizers have been major sources of phosphates that pollute our waters.

Students next transfer one milliliter of the nitrate or phosphate solution to a second liter of water. The concentration of the compound in the second container is now *one part per million*. If students were to transfer two milliliters of the solution to the second container, the concentration would be *two parts per million*; if students were to transfer three milliliters of the original solution to the second container of water, the concentration would be *three parts per million*. Because sixteen to twenty drops from a medicine dropper is about one milliliter in volume, a simple way to transfer a milliliter of the original solution into the second container is to squeeze in twenty drops. The student team preparing the test solutions should label each container, e.g., "Two parts per million of detergent X" or "Three parts per million of sodium phosphate."

After thoroughly mixing the resulting solution, students test it with the kit. Having done the measuring, diluting, and mixing themselves, they have a more concrete understanding of what is meant when the test kit results are expressed in parts per million.

Once students understand how to use the test kit, they can go on to investigate samples of water obtained from ponds and streams, comparing levels of phosphate in different samples. They can test the phosphate content of different detergents by making solutions of equal concentrations from the detergents using the dilution techniques outlined above and then test the resulting solutions using the kit. They might examine the level of phosphates or nitrates in the classroom aquarium. Is the aquarium in good condition? How does its nutrient level compare with the levels found in a local stream or lake?

When analyzing the data they obtained from their own investigations, students may find it helpful to compare their data to the levels that a number of states and the Province of Ontario allow in waste material being dumped into bodies of water by disposal plants. The level of phosphates and nitrates must be less than one part per million!

If you do not have liter containers and gram measures available, do not give up. Instead of using liter jars, students can use gallon containers of the type in which restaurants or delicatessens obtain salad dressing or pickles. They fill the gallon container with tap water and add a small pinch of a phosphate compound. To make further dilutions, they transfer with a medicine dropper a drop or two of the solution to a second gallon container of tap water.

One thing the class must have, however, is a test kit. Hach Chemical Company (Box 907, Ames, Iowa 50010) sells a nitrate test cube kit and a phosphate test cube kit for about $7 each.

If purchase of a test kit is beyond the classroom budget, students can still investigate the effects of phosphates or nitrates on algae and plankton animals. Teams set up gallon jar aquaria, all putting in the same predetermined amount of pond water. A solution with a predetermined concentration of phosphate or nitrate compound is added, but each team's solution has a progressively greater concentration of the compound. One team prepares a control aquarium to which no nutrients are added. Finally, each team adds some algae and plankton taken from the pond or lake. If you cannot catch small animals from a pond, try to obtain some *Daphnia,* also called water fleas, from an aquarium supply store. These shops sell this common water animal as live food for tropical fish.

The teams examine their jars every day and record any changes they observe. Is the number of plants and/or animals increasing? Is the water turning green with algae? Are the organisms dying? Is the water becoming foul? After several days of observing, teams can share their findings with the total class so that comparisons can be made. Is the growth in all the containers equal? Which containers show greatest growth and increase in number of organisms? Under what conditions do the organisms die?

A major advantage of this last activity is that students must check their experiment daily. This can be done as students enter the classroom, putting them to work as soon as they arrive.

Acid or Alkaline?

One of the simplest chemical tests young people can make is to test lake, pond, river, and brook water samples for levels of acidity or alkalinity. To do this students use pH Hydrion paper commonly available from science supply houses. The paper comes in thin rolls. Students tear off a strip about four centimeters long, dip it in the water sample, and match the color change of the paper to a color scale provided to find the degree of acidity or alkalinity—what is known as the pH.

On the pH scale seven represents neutral, below seven acid, and above seven alkaline. The extremes of the pH scale are 0 and 14, but most lakes that students will test have a pH in the range of six to nine. For example, Cayuga Lake, New York, has a pH range from 7.5–8.5; Lake Farrington, New Jersey, measures between 6–8; acid bog lakes range between 4 and 6. Waters receiving drainage from coal-bearing strata or volcanoes can go much lower. Contaminants added by industries can change the pH of a lake or pond. Therefore, when buying test paper, order one with a range of 4 to 9 on the pH scale. With older investigators, you might want to use a more accurate, liquid-type pH test kit that the Hach Chemical Company sells for about $7. It has a pH range of 6.5–8.5.

To become acquainted with the pH scale, young people can test the pH of distilled, rain, boiled, and tap water. Pure distilled water is neutral with a pH of seven. Rain water absorbs some carbon dioxide as it falls, which will generally make it acid with a pH of 4–5. Boiling rain water drives out the carbon dioxide and brings the pH closer to the neutral point of seven.

As a beginning, too, students can test the pH of household

liquids such as tea, lemon juice, white vinegar, and ammonia.

And How Do We Use Our Water Environment?

In *A Practical Guide to Water Quality of Streams* (U.S. Department of the Interior, 1969, p. 70), F. W. Kittrell lists eight categories of water use:

1. Municipal water supply
2. Industrial water supply
3. Agricultural water supply
   Domestic farm supply
   Irrigation
   Livestock watering
4. Recreation
   General
   Swimming, wading, skiing
   Boating
   Esthetic enjoyment
5. Propagation of fish and other aquatic life and wildlife
   Sport fishing
   Commercial fishing
   Fur trapping
6. Hydropower production
7. Navigation
8. Waste disposal

The Kittrell listing can serve as a checklist for analyzing the uses of a particular body of water in the community. Students studying a water environment can take the list with them as they investigate brooks, rivers, lakes, and ponds. Since the *Guide* is a government publication, the listing is not copyrighted and you can reproduce it for student use. We suggest that when you reproduce it you leave blank space between items so that students can record their objective observations directly on their own duplicated copies.

LEARNING MORE ABOUT WATER POLLUTION
Getting the Facts

Students investigating the quality of a local body of water and those who want to find out more about water pollution problems that are affecting the nation can consult a number of sources of information such as the following:

☐ The state pollution control agency—according to Kittrell, this is probably the most complete source of information about pollution problems in the state.

☐ The state fish and game commission—information about fishing grounds in the state and attempts to restock lakes and ponds with fish.

☐ The game warden—information about local conditions.

☐ The state health department—information on water supplies and water purification plants in the state.

☐ Municipal water department officials—information on local water quality estimates and complaints lodged by residents about local water problems.

☐ The Federal Water Pollution Control Administration—data on water quality studies that have been conducted.

☐ Interstate water pollution control agencies such as the Ohio River Valley Sanitation Commission and the Interstate Commission on the Potomac River Basin—information on interstate rivers.

☐ River development agencies such as the Tennessee Valley Authority—data on water quality studies conducted in the service area of the authority; also maps of the service area.

☐ U. S. Geological Survey (1200 South Eads St., Arlington, Va. 22202)—large-scale topographic maps giving details of local areas and costing about seventy-five cents each. Send for a free listing of maps available for your region.

☐ U. S. Fish and Wildlife Service—information on fish and fishing; also data on water quality studies bearing on fish.
☐ Professional environmental consulting companies—information on local environmental studies. Each issue of *Environmental Science and Technology* has a directory of such companies; examine the listing to find a company in your area that might be able to send out a speaker to the classroom to explain how environmental sampling and testing are done.

Students with specific concerns can send to these sources for information. Rather than writing to ask for "everything you can tell me about pollution," they should identify the specific kinds of information or material for which they are looking.

Representatives from a local or state agency can be excellent classroom speakers. For example, a representative from your state's fish and game commission may be able to discuss with students the conditions necessary to breed fish, to explain the temperature and food conditions required, the sun or shade requirements of the fish, the need for running water, and the varieties of fish being bred. Additionally, she or he may bring films to show the class and booklets to distribute. Of course, a class visit to a fish hatchery is not only a meaningful learning experience but fun as well.

The authors know of one teacher in New Jersey who managed to arrive with his class at a stream at the exact moment a truck from the fish hatchery was stocking the stream. On the spot the driver explained what he was doing and gave an impromptu talk about the variety of fish being used to stock that particular stream. Needless to say, the students were delighted.

So check listings in your local telephone directory. Students can look under county listings for references to park commissions, sewer authorities, and so forth, under state listings for environmental agencies and fish and game commissions, and under U. S. government listings. Such a check of the telephone directory may provide some unexpected contacts. For instance, we were pleasantly surprised to discover that our state has an environmental hot line for citizens to use to report environmental problems they discover and to get answers to their questions.

Students may be interested in investigating state and national laws as well as local rules regulating the dumping of wastes into bodies of water; they may wish to check related laws currently being considered by the legislature. To obtain this kind of information, they can write to the state or federal Environmental Protection Agency or to their state legislator requesting information on what is being done within the state to protect water environments.

When students in grades five and up learn that laws regulating dumping are pending before the legislature, they can obtain copies of the bills being considered, read them, become knowledgeable about related factors, and write letters to their legislators taking a stand for or against the pending legislation. Not only does this activity make students aware of one kind of possible action by citizens in a democracy but it is also more than likely that the legislator will answer the letter, which in itself can be a reward to young people involved in a study of environmental problems.

During local, state, and national elections, young people studying environmental problems can analyze the statements of candidates to find out their position on environmental issues. For candidates already in office, students can check past voting records to see if their words reflect their actions. Often the League of Women Voters has information on the voting records of legislators.

Developing Case Study Reports

Much has been written about water pollution problems that have affected specific communities and particular bodies of water. For example, the League of Women Voters Education Fund in 1966 issued *The Big Water Fight* (Brattleboro, Vt.: Greene Press), which was billed as a discussion of the "Trials and Triumphs in Citizen Action of Problems of Supply, Pollution, Floods, and Planning Across the U.S.A." This general book describes numbers of specific situations; it would be useful to an upper-grade student who is writing a case study of the problems in one community.

In contrast, Leonard Stevens' *The Town That Launders Its Water* (Coward, McCann, 1971) zeroes in on how one community—Santee, California—is using reclaimed sewage water to supply five freshwater lakes that serve the community for recreational purposes. Good for students in grades five and above, the book is a fascinating account of how one town in an area where the average rainfall is just 25 centimeters per year solved its water problems.

Articles in popular magazines also provide information about specific water pollution problems. The August 1973 issue of *National Geographic* contains an article with full-color photographs called "The Great Lakes: Is It Too Late?" by Gordon Young, James Amos, and Martin Rogers. This well-written piece describes some of the pollution problems that threaten the survival of the United States' inland seas. An article in the *Natural History Magazine* (November 1968) by Arthur Hasler and Bruce Ingersoll, "Dwindling Lakes," supplies additional information on the pollution situation in the Great Lakes. An upper-grade student doing a case study of Lake Erie and the other Great Lakes will find these articles as well as "Life in a Dying Lake" by Peter Schrag in *Saturday Review* (September 20, 1969) tremendously helpful. In *Fortune* (February 1970) "The Limited War on Water Pollution" by Gene Bylinsky tells about the 1969 situation on the Cuyahoga River in Cleveland when the water became so laden with oily wastes that it burst into flames and destroyed two railroad bridges. In the

National Geographic, "Our Ecological Crisis" (December 1970) shows a full-color photograph of the Cuyahoga and goes on to describe other aspects of the crisis that faces our environment. Both articles are superior references for upper-grade students involved in case-study reporting.

The articles noted above are only examples of the kinds of materials students will find in journals. Sixth grade is not too soon to introduce young people to the *Reader's Guide to Periodical Literature,* a standard reference in the public library. Using the *Guide,* students can begin to locate appropriate source materials on their own.

And don't forget newspapers. As citizens become more aware of environmental problems facing them, it is not at all uncommon to find news reports and feature articles that can be helpful to students preparing reports of specific environmental problems.

Filmed materials also can supply background information for case studies. Several films describe the Great Lakes' crisis. Encyclopaedia Britannica Films has *The Aging of Lakes,* a fourteen-minute, color film that explores conditions leading to eutrophication. An eleven-minute, color film, *The Great Lakes: A Matter of Survival,* available through BFA Educational Media (Santa Monica, Calif.), tells what can be done and is being done to save the lakes. Both films make excellent viewing for students in grades five and up.

Men at Bay, a twenty-six-minute color film endorsed by the Earth Science Curriculum Project, explores the "death" of a bay. The documentary is suitable for students in grades six and above. *The Drowning Bay* is a shortened version of this documentary. Suitable for students in primary and middle elementary grades, it illustrates what can happen when people allow a body of water to die from pollution. Both films are available on a rental basis from Modern Film Rentals (Summit, N.J.; Elk Grove Village, Ill.; Los Angeles; and Atlanta).

To Eat Clams or Not to Eat Clams?

That people are affected by contaminants in the water through the food that they eat has unfortunately been demonstrated many times. One of the best-known occurrences was the disaster at Minimata Bay, Japan. Industrial wastes high in mercury were dumped into the bay waters by an acetaldehyde factory. Fish living in the bay assimilated the mercury through food they ingested. People in turn ate the fish. The result—44 people dead from mercury poisoning between 1953 and 1961. Boys and girls may remember a similar mercury scare in the United States in which tuna fish high in mercury were discovered and many people stopped eating tuna. They may have heard of fish kills on the Mississippi River where vast numbers of fish were killed by excessive and toxic wastes dumped into the waters.

Obviously it is impossible for young people to study firsthand situations such as these unless by chance they live on a bay, river, or gulf where fish are actually dying. However, the topic is an interesting one for researching and reporting. Students can read to find out about:

☐ Sewage, shellfish, and hepatitis: A student might decide to call his or her report "To Eat Oysters or Not to Eat Oysters, That Is the Question."
☐ Mercury, fish, and human poisoning: A student might title the report "Can We Prevent Future Minimatas?"
☐ Pesticides, fish, and human poisoning: This report could be called "DDT—What Do We Do About It?"
☐ Radioactive elements in the food chain: "Fishy Phosphorus 32" would be a good title.

To find out about contaminants in the sea, students should consider that fish and other marine life concentrate pollutants, "acting like filters, or sponges, to soak up tiny amounts until they become appreciable" (D. S. Halacy, *Now or Never: The*

Fight Against Pollution, Four Winds Press, 1971, p. 81). *Now or Never* develops this point effectively and is an in-depth reference for students in grades five through eight who investigate the effects of contaminants on humans. The chapter on pesticides is particularly valuable as it describes some of the problems of DDT and ways humans come into contact with the pesticide. If you get into a study of the effects of pesticides on life, remember the classic—Rachel Carson's *Silent Spring* (rev. ed., Fawcett World, 1973). It is both fascinating and frightening! And for recreational reading, introduce students to Jean Craighead George's *Who Really Killed Cock Robin?* (Dutton, 1971). The answer to the question posed in the title of this detective story for young people in grades four to eight is pollution.

There is one firsthand observation that students from fifth grade and up can undertake—a study of the structure and eating habits of the clam or oyster. Clams and oysters are filter feeders that live in mud. Both siphon water through their bodies, extracting nutrients in the process. Because the clam and oyster

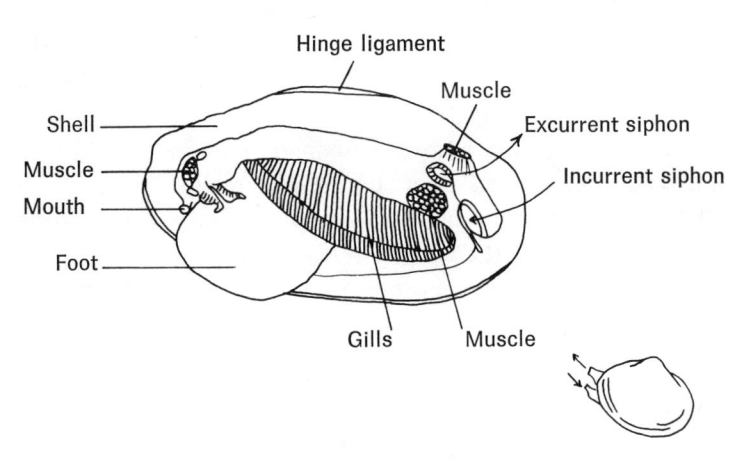

ANODONTA, THE FRESHWATER CLAM

live in the muds of gulfs and bays, they may well siphon contaminated water through their bodies, retain the contaminants, and pass them on to people who eat shellfish.

Clams can be purchased from a seafood market, or in suitable areas, students may dig clams to bring in samples. Using the diagram of the internal structure of the clam as a guide, students dissect the animal, identify the incurrent and excurrent siphons, and see the fleshy portion in which contaminants may be retained. If a live clam can be found and there is an aquarium in which to put it, you may wish to inject some food coloring or plain carmine stain near it. Students can watch as the clam takes the dye into its body through the incurrent siphon and ejects it later from the excurrent siphon.

This Is Our Country

As students read and discover facts about pollution in the rivers and lakes of their country, they can make a large-scale map of the nation, indicate its major bodies of water, and plot their data directly on it in a visual way.

Use an opaque projector to project a map on a bulletin board which has been covered with light green construction paper. The ocean boundaries as well as the lakes and rivers are traced in blue; boundaries with other nations are indicated in black; mountain ranges are traced in brown. The resulting outline map will be larger than the original and fill the entire bulletin board if the opaque projector is focused just right.

Next students make tiny red flags from construction paper and attach them to the ends of straight pins. As they discover in their reading that a particular body of water, e.g., the Hudson River, Lake Erie, or the Mississippi, has a high pollution level, they stick a flag pin in the approximate location on the map. As flags begin to appear in larger and larger numbers, students will begin to conceive the magnitude of the pollution problem facing their country.

CHAPTER 5

DENUDING THE FIELDS
Activities with Soil and Pavement

We have become rich through the lavish use of our natural resources, and we have just reason to be proud of our growth. But the time has come to inquire seriously what will happen when our forests are gone, when the coal, the iron, the oil and gas are exhausted, when the soil has been further impoverished and washed into the streams, polluting the rivers, denuding the fields and obstructing navigation.

> Theodore Roosevelt,
> Conference on the Conservation of
> Natural Resources, 1908

In "civilized" countries today vast expanses of pavement are taking the place of fields and forests as shopping centers, airports, freeways, houses, and apartments are added to the landscape. Each new development increases the amount of pavement, the destruction of trees and grasses, and ultimately the amount of water that runs off the land. Increased runoff decreases the water available in the soil for plant growth. It means that there is a greater possibility of flooding and that the runoff will have greater force—a force that is a factor in erosion.

In "civilized" countries, too, people are at work planting to feed the growing populations of the world and mining natural resources to supply rising energy needs. To meet food needs, farmers strip the natural cover of grasses and trees from the land, exposing the soil to the forces of erosion—wind, water and gravity. To meet energy needs, miners often do the same, leaving miles of exposed and minerally depleted land behind the consuming mouth of the excavation scoop.

In short, use of the land can create immediate problems, many of which students can investigate.

INVESTIGATING SOIL-WATER RELATIONSHIPS

The Water Cycle

Study of the water cycle provides young people with a framework for investigating water-soil relationships. Working from a series of observations and experiments, they can put together the "pieces of the puzzle" and construct a schematic similar to the one shown in the diagram of the hydrologic cycle in which water travels from ocean to sky to land and back to the ocean.

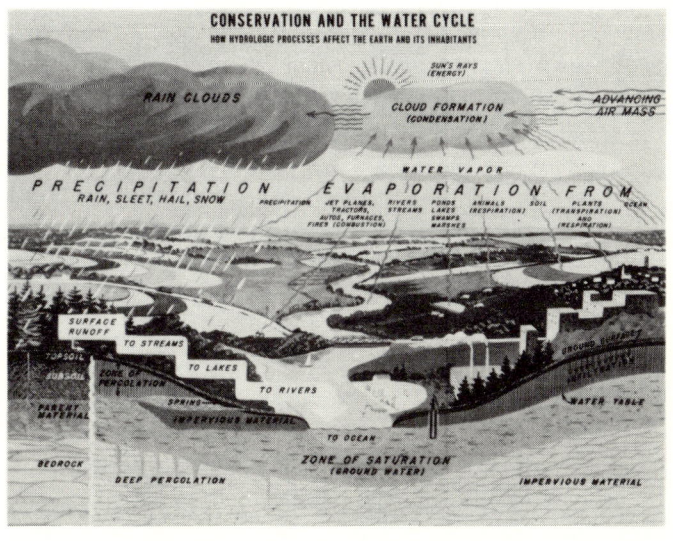

Reprinted with permission of Soil Conservation Service, U.S. Department of Agriculture from *Conservation and the Water Cycle*, 1972.

An easy beginning for building an understanding of the water cycle is to focus on one process—evaporation. Students working in investigative teams put small, measured amounts of water into shallow vessels such as petri dishes or aluminum pie plates. They put the plates outdoors under different conditions, e.g., windy, calm, humid, dry, hot, and cold days and in direct sunlight and shade. They check to see how long it takes under each condition for the water to evaporate.

Young people living in cities can check the time it takes for water puddles on street pavements to evaporate after a rain storm; they can compare evaporation times after different storms. If the schoolyard has a paved area, they can sprinkle it with water and check evaporation time. Students record their data in tabular form; in addition, they place arrows labeled *evaporation* on the beginning of a water cycle chart previously outlined on a bulletin board by a student.

Additional upward-directed arrows are drawn on the bulletin board chart after students have observed the water given off into the air by plants through what is known as transpiration. Students wrap a plastic bag around a pot of a leafy, adequately watered house plant so that water from the soil and the surface of the pot cannot escape by evaporation. Then they place the leafy plant and pot into a second, larger plastic bag and close the bag. After the plant has stood in sunlight for fifteen or

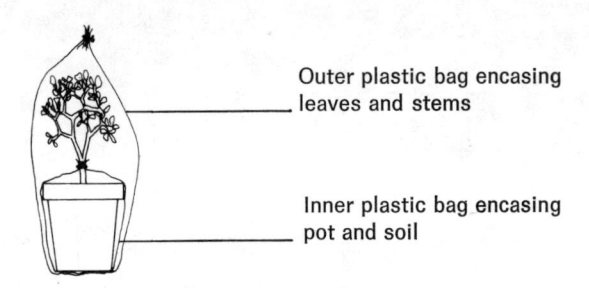

Outer plastic bag encasing leaves and stems

Inner plastic bag encasing pot and soil

TRANSPIRATION DEMONSTRATION

twenty minutes, they should be able to see a film of moisture collecting on the inside of the plastic bag covering the plant and sometimes a second film of moisture on the inside of the plastic bag encasing the pot and soil. They can add symbols for vegetation and arrows marked transpiration to their chart.

It can be demonstrated that animals release moisture into the air as well. Ask five students to hold their mouths close to the chalkboard and exhale onto the board. Moisture spots will appear, which will soon disappear as the water evaporates. Student participants can sketch symbols for animal life and add evaporation arrows onto their water cycle chart.

The precipitation part of the water cycle puzzle can be shown in a classic demonstration. Rapidly boil some water using a hot plate as the source of heat. Wearing protective gloves, hold a metal tray on which there are some ice cubes over the boiling water. As the water vapor strikes the bottom of the cold tray, it liquifies and as the liquid collects, it trickles down as "rain." Rain clouds and downward arrows marked precipitation are added to the bulletin board diagram by students who have talked about this demonstration.

The runoff-infiltration part of the cycle can be demonstrated and investigated in a number of different ways. Students can dig a hole in the ground about fifteen centimeters deep. They first feel the soil taken from the hole for wetness. Next they fill the hole with water, wait until most of the water has seeped out of the hole, and dig down *next to the hole,* feeling the soil for wetness. They will discover that not only the soil beneath the hole is wet, but that the soil next to the hole is wet from the water that has infiltrated laterally.

Students can fill a flat box with soil, prop the box at a 30° angle, and sprinkle a measured amount of water across the top of the box. At the bottom they collect the water that runs directly off the soil in a trough or flat pan and subtract that amount from the amount originally sprinkled on the soil. The difference is the water infiltrating the soil. They may wish to repeat the experiment using fresh soil each time to find out how the slope of the box affects the amount of water that runs off and infiltrates. They can finally observe runoff during and after storms of differing magnitude. Is there a relationship between the amount of runoff and the intensity of the storm? After observing runoff and infiltration, students can complete their water cycle schematic with arrows representing runoff and infiltration.

Will Water Penetrate?

Soils are comprised of particles that range in size from very coarse sands to extremely fine clays. How much of each type a specific soil contains and how closely together the particles are packed influence the rate at which water infiltrates between the particles and penetrates the soil. The relationships between particle size and water penetration and between closeness of particles and water penetration are easily investigated by students in grades four and above.

Divide the class into three investigative teams with each responsible for one of the following activities. Each team carefully labels the components of its experiment to aid in the explanation it will give to the total class.

Members of one team bring in discarded soup cans with both ends cut out and the rim of one end removed leaving a sharp edge. They push the sharp-ended cans several centimeters into the ground at different locations, add equal measured volumes of water, and record the time it takes the water to penetrate the soil. They examine the soil around the cans to see what kinds of particles are present and how tightly they are packed. If different soils—woodsy, cultivated, grassy, sandy, or clay—are not present near the school, individual students can carry out the tests in home areas and bring the data and a sample of the soil used to class.

A second team carries out a variation of the water penetration experiment across and adjacent to a well-traversed path. First numbering the cans to aid in data recording, team members press several sharp-ended cans into the path and the area nearby. Simultaneously they pour equal volumes of water into the cans and in turn call out descriptions about how fast each can is emptying—if at all. One member records the data in tabular form:

WATER PENETRATION TABLE			
SITES	AFTER 30 SECONDS	AFTER 1 MINUTE	AFTER 5 MINUTES
Center of the path			
Half-meter from the center of the path			
Edge of the path			
Half-meter from the path			
Two meters from the path			

After performing the experiment, team members speculate on reasons that could explain their findings.

A third team uses worn-out ballpoint pens or pencils to study penetration. They push their pens or pencils into the ground at locations across and adjacent to a well-traversed path and on grassy, cultivated, and uncultivated sites. They record the depths to which the pens penetrate before the pressure makes the investigators' fingers feel uncomfortable and further penetration impossible.

Students who have conducted penetration studies out-of-doors may wish to attempt more systematic inside investigations

Denuding the Fields

with soils of differing textures. The texture of a soil depends on the particular combination of sizes of the sand, silt, and clay particles in that soil. Particle sizes are shown in the table:

SIZES OF PARTICLES IN THE SOIL

PARTICLE	DIAMETER RANGE IN MILLIMETERS
Very coarse sand	2.0 –1.0
Coarse sand	1.0 – .5
Medium sand	.5 – .25
Fine sand	.25– .10
Very fine sand	.10– .05
Silt	.05– .002
Clay	below .002

If you have access to sand, silt, and clay, students can make soils of differing particle sizes to test for penetration rate.

A relatively simple test is made using glass lamp chimneys or other transparent tubes, stretching cheesecloth or pieces of nylon stocking over the bottom ends, and fastening them with elastic bands. The tubes are partially filled to the same levels with test soils—dry coarse sand, fine sand, silt, clay, a mixture of clay and sand, a mixture of clay and silt, or local soil—without packing down the soil or leaving air pockets. The glass chimneys or columns are supported above collecting jars by ring stands and clamps or by any other method students can devise. Finally, a measured volume of water is added to each column. Students should be ready to time the passage of the water through the sand column immediately; in contrast, water may not move through the clay column for a number of hours.

Students can observe and record the distance the water travels downward in thirty seconds; a strip of tape marked off in centimeters and pasted vertically to the column facilitates measurement. Data from all the soil columns are compared to determine the relationship between penetration rate and soil texture.

Students can also observe and record the volume of water draining through and being caught in the jar beneath. The difference between the amount of water that enters the soil at the top and what collects in the jar beneath is the amount retained by the soil particles. These data are used to calculate the water retention ability of each type of soil texture. For example, if 30 milliliters of water entered the soil at the top and 25 milliliters of water were collected, 5 mililiters were retained. Thus the ratio retained is one-sixth. If some water is still standing on top of the clay when measurements are taken, students draw off the water with a medicine dropper and subtract that volume from the amount originally added to the soil.

Measuring Raindrop Splash Power

Raindrops do not hit the ground gently. On the contrary, they bombard soil with a velocity of about 24 kilometers per hour (15 mph). Each drop throws up a tiny crownlike film of water around the impact point and lifts soil particles from the surface. Once soil particles are lifted by raindrops, a very small wind or water flow will carry them away. This is how erosion begins.

Student investigators can study raindrop splash power by making portable splash recorders. They nail boards about 10 x 30 centimeters at right angles to other boards to form bookend-like structures. They paint the faces of the uprights white or tack on white cardboard to pick up the mud splashes, nail narrow strips of wood or metal along the tops as overhangs to keep rain from running down the faces of the uprights, and place rocks on the support boards to hold the upright splash faces firm. They place their splash recorders on bare hard soil, cultivated soil, or grassy area and after a rain, they compare their splash records to find the degree of soil lifting caused by the raindrops.

Follow-up questions for students to discuss are: How high did

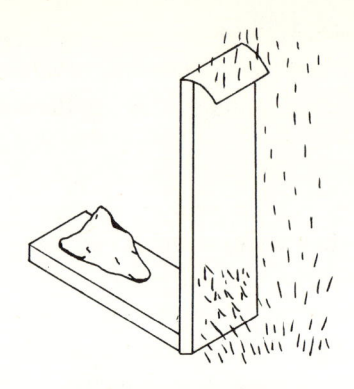

SPLASHBOARD

the splashes reach? Where did splashboards receive the most soil—in grassy areas? in bare soil? in wooded areas? What is the value of a plant cover in protecting soil from erosion? When during the year would erosion on cultivated fields be the greatest? Would fields of row crops such as corn, or fields of grasses and grain crops be subject to the greatest erosion?

To demonstrate use of the splashboard, one may be set up outside and sprinkled with water from above with a watering can. This, however, is not so realistic and effective as having students see the results of soil thrown up by actual raindrops.

A variation can be constructed to make splash distance determinations. Obtain large pieces of corrugated cardboard by breaking up supply cartons furnished by the custodian. Cut out and discard a 20-centimeter diameter circle from the center of each piece of cardboard. When rain is expected, place the corrugated boards outside on bare and cultivated plots, in grassy, and on cultivated areas mulched with straw. If it is windy, place bricks or rocks on the corners of the mud catchers. After the rain, students push a sharpened pencil into the center of the hole and with a meter stick measure the distance from the pencil that soil splashed onto the cardboard.

They can consider whether soil was splashed outwardly different distances under different conditions of soil cultivation and plant growth and whether the amount of splashed soil changed as the distance from the center increased.

Is There a Crust?

Raindrop bombardment tends to make the fine particles of soil fill the spaces between the larger granules, causing the soil surface gradually to develop a crust. When this happens, instead of soaking into the soil, water puddles or runs off if the surface is sloped.

Crusts of over 2.5 centimeters are not uncommon and can seriously affect the amount of water retained by the soil. One research study in Coshocton, Ohio, indicated that only 1.27 centimeters or about half of a one-inch July rainstorm was absorbed by a crusted cornland soil. The large runoff carried two tons or 1,814.4 kilograms of soil off each acre (.405 hectare) of this cornfield. Records showed that without the crust, the soil could have absorbed nearly three times the water dropped by that rainstorm.

Students—especially those living in farming areas—may wish to demonstrate the effect of crust formation on water penetration. The procedure is 1) Set up two funnels and half fill each with the same kind of soil, taking care not to pack the soil down; 2) Cover the surface of one funnel with a centimeter or two of short pieces of straw, dried grass, or wood shavings; 3) Place a jar under each funnel spout; 4) Gently sprinkle both funnels with water; 5) Observe whether water flows more readily through the bare or the protected soil; 6) After all the water has penetrated the soil, gently touch the surface of the bare soil to see if the beginnings of a crust can be felt.

Runoff and Erosion

After a storm, encourage students to bring in samples of

runoff water from different locations. Samples that are slightly muddy can be filtered through filter paper. If all students use samples of equal volume and record conditions under which they were taken, they can visually compare the amounts of sediment collected and hypothesize about the conditions under which runoff water carries a lot of suspended material.

Students should allow very muddy samples to stand until the eroded material settles out. They can compare the gradual buildup of sediments at the bottoms of the containers.

An observational walk on a day after a heavy rainstorm will impress on students the fact that water rapidly running off the land carries soil and other materials with it and deposits them in lower areas. Along city streets, look for places in higher areas where rushing water has literally washed parts of the streets clean; look for locations in lower areas where some of this material has been deposited as the water slowed down. In a more rural area students can study the floor of a stream; there they will find boulder fragments carried by the stream when it was filled with water after a heavy rain. They will find areas along broad meanders where very fine sand has been deposited as the stream slowed down almost to a crawl. Or students can examine the edge of a farm field under cultivation for ruts and deposited materials.

Students of all ages will enjoy such observational excursions. With younger boys and girls schedule a talk-time upon returning to the classroom so that children can orally describe their observations.

Young people who live along a major river can also investigate the effects that material deposited in a channel can have on shipping. Such an investigation makes an interesting researching-reporting study for an individual student who may document a written or oral report with photographs of sediments deposited in the local river. Sometimes sandbars and silt buildups along wide meanders are easily observable to the searching eye of reporter and camera.

In their classrooms, students can simulate erosion conditions by constructing erosion boxes. Obtain several plant flats of the kind garden shops or nurseries use to grow seedlings or to carry small potted plants; shallow wooden fruit boxes can also be used. Line the flats or boxes with plastic sheeting to waterproof them; several layers of dry cleaners' clothes bags serve well as a lining. Put soil in a box, tamp it down, and sow grass seed. Before watering, firm the grass seed in the soil by tamping it with a block of wood. Or purchase or cut a few pieces of sod as "instant grass" for immediate experimental use; in this case, cut the sod to fit an erosion box and place it above a soil layer already placed in the box. To simulate row crops, plant bean, corn, sorghum, tomato, or sunflower seeds in other flats.

On the day seeds are sown, also prepare one or more boxes of soil with no seeds. Gently water these boxes periodically so that the soil particles will settle in as will happen in the seeded boxes.

When plant growth is well established, students can use the boxes to test the soil-holding ability of vegetation. Work with them first to establish what information they are seeking and how the experimental tests will be conducted. Through discussion students will develop lists of questions to which they are seeking answers. Lists might contain such questions as: Will gentle sprinkling lift and move more soil from a grassed area? an area planted with row crops? a bare soil area? If the boxes are raised on one end to make a sloping surface, how will that change the amount of erosion? Are row-cropped, grassy, and bare soil areas similarly affected by the increased slope of the land? Does it make any difference whether the box is tilted so the rows run across or up and down?

Plant flats, sprinkling cans, and water may have to be carried outside to conduct the actual experiments unless there is some way to keep mud out of the sink and off the classroom floor. This activity can become really messy and the custodian will probably not appreciate the big clean-up job. Another advantage of moving outside is that a garden hose may be used to spray water on the prepared erosion boxes.

After the experiments have been run and results recorded, ask students to consider the ways the experiments duplicated natural conditions and the ways the experiments may be limited or produce unrealistic results. For example, the soil in the unplanted box may not be as stable as the natural soil outside; furthermore, the depth of the soil in the boxes is less than that under grass or crops outside, and the soil may thus become waterlogged or slip on its plastic sheeting. California students may recognize such slippage as comparable to mud-slide conditions with which they are acquainted.

Students can also make on-the-spot surveys of local erosion sites. Essentially investigators ask the same questions as were asked during experimentation with the erosion boxes, e.g., Did previous runoff remove as much soil from a grassy area as from a bare soil area? from a grassy area as from a row-crop area? How does erosion in highly sloped areas compare to that on more level areas? What evidences are there that people are attempting to halt erosion caused by flowing water?

The class can go on to select a small local area showing moderate or potential erosion damage and attempt to repair the damage or prevent future erosion. They can fill small ruts with soil, fertilize the land, and plant grass seed, or they can cultivate the soil on bare, sloping surfaces, fertilize, and sow seed. In either case, students should cover the seeded area with salt hay or burlap pinned down with stakes or rocks. The burlap or hay holds the soil and seeds until the grass grows through it to a height of several centimeters. At that point the burlap can be carefully peeled away.

If the class wishes to attempt larger erosion repair work, the county agricultural agent can assist with advice and can furnish soil conservation pamphlets that will aid student planning of such projects.

INVESTIGATING PAVEMENT–WATER RELATIONSHIPS

Paving the World

In a U.S. Geological Survey Circular, *Hydrology for Urban Land Planning—A Guidebook on the Hydrologic Effects of Urban Land Use* (1968), Luna B. Leopold asserts that, "The volume of runoff is governed primarily by infiltration characteristics and is related to land slope and soil type as well as to the type of vegetation cover. It is thus directly related to the percentage of the area covered by roofs, streets, and other impervious surfaces . . . ." Studies of the effects of urban development on runoff volume supply such astounding facts as that a home built on a 6,000 square foot plot renders 80 percent of the lot impervious to water and that, "Sediment derived by erosion from an acre of ground under construction in developments and highways may exceed 20,000 to 40,000 times the amount eroded from farms and woodlands in an equivalent period of time."

Because young people living both in urban and suburban areas are being surrounded increasingly by roofs and streets and decreasingly by trees and grass, a pavement-water study is particularly relevant. An easy beginning is to test the imperviousness of pavement. Select a sloped area of the schoolyard and pour water over it; students will see how the water moves quickly into lower areas, leaving a film behind to evaporate into the air, and will conclude that almost 100 percent of rainfall becomes runoff. Encourage students, especially young observers, to talk about what they have seen.

Students who have witnessed this simple demonstration can go on to calculate the volume of runoff from a paved schoolyard when a rain of one inch falls on the yard. To do so they measure and multiply length by width to obtain the area in square feet. They multiply by one twelfth to get the cubic feet of runoff produced by a one-inch rainfall. Now when students listen to weather reports and hear that the rainfall yesterday was one-half inch, two inches, or even five inches, they can quickly calculate the cubic feet of runoff from the schoolyard. As long as American weather reports announce rainfall in inches, it is probably simpler to use the traditional system of measurement in making calculations; however, where rainfall is announced in centimeters, the area of the schoolyard should be calculated in square meters and the volume of runoff computed in cubic meters.

Having observed the volume of runoff from a small impervious plot, students can project some of the negative results of an almost 100 percent runoff. Points they may project include: less water available in the soil for plant growth and a possible lowering of the water table, increased chances of flooding as runoff water moves to lower areas quickly after a rain, destruction of stream channels as large volumes of runoff hit a stream simultaneously, and increased erosion as runoff water moves forcefully across the land.

A next step to consider is the amount of area in the neighborhood covered by roofs and pavements. In residential areas, students can ask their parents for the dimensions of their plots and the square footage of land-occupying parts of the house—basically the roof area. They can measure or estimate the widths and lengths of slate, concrete, or hardtop paths and drives and can calculate the percent of their land covered by impervious surfaces. The formula for calculating percent of land covered by impervious surfaces is:

$$\frac{\text{area of roofs + walks + drives}}{\text{total area of lot}} \times 100$$

In urban areas where high-rise apartment buildings and multi-family dwellings are the norm, students can list all the impervious surfaces in the neighborhood plus areas of grass, trees, and soil. Their listings can serve as a base for making a rough estimate of the percentage of land covered by impervious surfaces.

If all the students in a residential section calculate the area covered by impervious surfaces on their home sites, they can compute the volume of runoff from their combined home sites after different amounts of rainfall. They can estimate the area covered by a nearby shopping center, industrial park or apartment building complex, estimate the area covered by impervious surfaces, and compute runoff after rains of known size.

As another project students can interview residents living in low areas or near any development that has contributed large impervious surfaces such as shopping centers, highways, airports, apartment complexes, or industrial centers. They can ask such questions as: During rains has there been flooding? If so, from where did the water come? Has the flooding increased in recent years? If so, why? They can also visit construction sites where excavation has removed plant cover, look for signs of erosion, and ask construction workers what is being done to prevent soil wash and unnecessary destruction of trees.

Pavement vs. Grass

Once young people in urban and suburban areas have become involved in pavement-water studies, they may wish to consider a side effect of large asphalt or concrete surfaces—temperature increase. Everyone has probably felt heat reflected from a hot pavement and enjoyed walking barefoot through cool, lush grass. An activity can be based on this difference in the heat-holding ability of pavement and grass.

Students start by selecting a location that has contrasting surfaces such as bare soil, gravel, blacktop, concrete pavement, and grass near one another and by placing plain stem thermometers on each surface. If the thermometers have the degree markings on a cardboard or wooden back, the part behind the bulb can be cut away so that when the thermometers are used

Denuding the Fields

there is nothing between the bulb and the surface. To make fair comparisons, students should place thermometers in the same manner at each test site; that is, one bulb should not lie on the surface while another is buried under soil or gravel. Thermometers should be left in position for at least five minutes before students take readings. During the waiting period, they can jot in their field notebooks a three- or four-word description of each experimental site, leaving a space for each reading. By the way, the same surfaces can be checked for differences in temperature when exposed to sun or shade.

Before leaving the sites and returning to their classroom, investigators rest their hands on test surfaces and then hold their hands a very short distance from them. The latter detects the heat that is conveyed through convection currents in the air without direct contact with the surface.

On a warm day experimenters will probably find a difference of over five degrees between sun and shade and several more degrees between grass and pavement. The experiment can be repeated on a cold day, a cloudy day, and even on a rainy day when the thermometers get wet. Follow-up questions to relate temperature difference to comfort include: What comfort differences might we feel if we stand on a pavement in the sun or on a grassy place nearby? What part does heat reflection from a surface play in the temperature differences we feel? Could you feel with your hand the temperature differences the thermometers showed? What differences might grassy or shady areas have in making city living more comfortable? Would the same differences be desirable both winter and summer?

MAINTAINING SOIL PRODUCTIVITY

Looking for Soil Profiles

In the previous section, students found that soils are mixtures of different-sized particles that are derived from fragments of weathered rocks and minerals. In addition, there are varying amounts of organic matter in soil and air and water are found in spaces between particles. These components of soil are not randomly arranged; rather, climate and living organisms act on soils to develop layers or *horizons.* A cross-sectional view of soil horizons from the surface downward is known as *the soil profile.*

If a road cuts through a hillside or an excavation for a commercial building or house is near the school, the class can visit and observe the soil profile. Suggest to class members that on their way to school they look for a profile where the horizons are visible. When the class visits, one or two students should bring along a small shovel or hoe to expose a fresh, clear, vertical surface if fallen debris or erosion has obscured the profile.

If ready-made profiles are not located near the school, young people can share the labor of digging a pit about a half meter long, a meter deep, and wide enough to get a view of the exposed vertical surface. After observations have been made, other students fill in the pit to restore the land and prevent possible tripping accidents.

If it is impossible to show students the real thing, make a transparency from the profile in the accompanying diagram and project it with an overhead projector. Students can tell about similar profiles they have seen as they have driven or walked along highway or construction sites.

Interpreting Soil Profiles

In interpreting soil profiles with students, be aware of the general horizons or layers to look for and of possible regional or local variations that may be encountered.

A profile in an undisturbed woodland or grassland may show several horizons above bedrock. The two most easily discernable are what gardeners call topsoil and subsoil. The thickness, color, texture, particle structure, and other features of these horizons

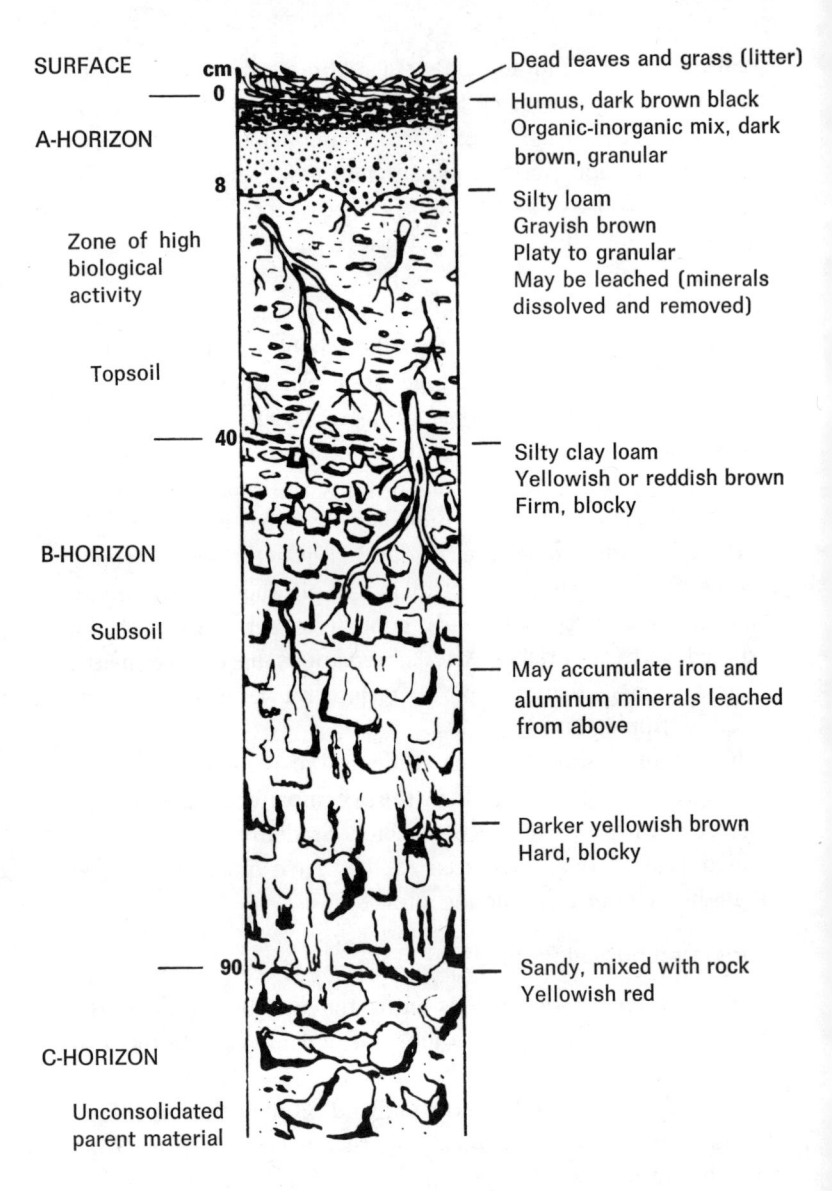

	cm		
SURFACE			Dead leaves and grass (litter)
	0		Humus, dark brown black
A-HORIZON			Organic-inorganic mix, dark brown, granular
	8		Silty loam
Zone of high biological activity			Grayish brown
			Platy to granular
			May be leached (minerals dissolved and removed)
Topsoil			
	40		Silty clay loam
			Yellowish or reddish brown
			Firm, blocky
B-HORIZON			
Subsoil			May accumulate iron and aluminum minerals leached from above
			Darker yellowish brown
			Hard, blocky
	90		Sandy, mixed with rock
			Yellowish red
C-HORIZON			
Unconsolidated parent material			

A GENERALIZED SOIL PROFILE

vary from one geographic area to another and determine the agricultural value of the soil.

In temperate, humid regions, one may expect to find that the uppermost soil layer just below the ground surface tends to be dark because of organic matter from decaying vegetation and leaves mixed with the mineral matter. Just below that, a lighter, sometimes grayish zone may exist because rainwater soaking through has dissolved out and carried away some minerals and deposited them in a lower horizon. When these minerals reach the subsoil, or B-horizon, many are dropped and accumulate, making the horizon appear reddish or yellowish brown with a clay texture. The C-horizon is made of disintegrating particles and pieces of broken rock. It is sometimes called the parent material.

In less humid grasslands, the topsoil may be very thick—possibly more than half a meter—and the subsoil may show white limy concretions. In desert areas, the profile may consist of an upper horizon of granular mineral matter with little organic matter and a lower one of disintegrated rock.

As you and your class examine the horizons of a soil profile, encourage students to describe each horizon in their notebooks using the characteristics outlined below:

Color: Look at each layer. Is it black from organic matter? red or yellow from iron and aluminum minerals? whitish or grayish from mineral loss or limy deposits? bluish, grayish, or mottled with yellow and blue because of poor drainage?

Texture: Pick up about a teaspoonful of material from each layer and rub the soil between thumb and forefinger. Then add a few drops of water to the soil, knead it, and again press the soil between forefinger and thumb. Is it gritty, loose, and not sticky when moist? If so, it is probably a sandy soil. Does it feel smooth and make a slick, sticky ribbon when wet? Does it form hard clods when dry? If so, it can be identified as a clayey soil. Does it feel smooth and powdery when dry and make a smooth ribbon that is not sticky when wet? If so, it can be classed as a silty soil.

Of course, you and your students cannot generally expect to find horizons comprised of 100 percent sand, silt, or clay. Soils have different proportions of these components. Sandy soils are over 70 percent sand; clayey soils are at least 35 percent clay. About midway between sands and clays are the loams. Loamy soils are moderately coarse to moderately fine textured.

An analysis of soil texture leads to an evaluation of the poorness or goodness of the soil. A poor soil for plant growth dries out rapidly; it has too many large particles, such as sand, that allow water to drain too easily. Also a soil is rated poor if it has too many clay particles that are so fine and tightly packed that excess water does not drain. Conversely, a good soil has a mixture of particles that allows adequate drainage.

Organic Matter: Examine the soil for the presence of organic matter. Usually the uppermost horizons contain the greatest amount and have high biological activity; organic matter increases the water-holding capacity of sandy soils and reduces the stickiness of clayey soils. Students can use the following question sequence to guide their examination of the soil:

Is organic matter being broken down into humus at the surface of the soil? Are there fragments of leaves, twigs, or other plant parts in the process of becoming humus? Is the humus becoming incorporated into the topsoil or A-horizon? Organic matter becoming part of the soil is nature's way of recycling some of the minerals that plants have removed during growth.

How far can plant roots penetrate the soil profile? What evidence is there to support your hypothesis? Do the roots penetrate the B-horizon? Can air from the surface get to plant roots?

Are earthworms or ants present? Is there evidence that they are mixing the soil? Is there any evidence that organic matter is affecting the water-holding ability of the soil?

Structure: Examine the soil for structure—the arrangement of soil particles into aggregates. Factors influencing structure include rainfall, amount of organic matter, presence of microorganisms, and farm tillage methods. Using the table below as

CHART FOR INTERPRETING SOIL STRUCTURES

TYPE OF STRUCTURE	APPEARANCE	DESCRIPTION	HORIZON WHERE TYPE OCCURS
Granular		Loose, grainy particles	A and B
Prismatic		Vertical, angular, prismlike aggregates; often 13 centimeters long	B
Columnar		Pillarlike aggregates with rounded edges and tops	B
Blocky		Crude, cubelike aggregates, stacked like blocks; about 2–3 centimeters on a side	B
Platy		Thin, crude aggregates, platelike or flaky, horizontally placed	A and B

Denuding the Fields 131

a guide, students can describe the structure of the soil layers in the profile they are examining. Suggest that they use the adjectives given in the chart as they write their descriptions.

Plant and animal debris—including dead leaves and insect remains—are converted to humus and mineral matter by the metabolism of microorganisms and many near-microscopic animals that feed on the debris and on each other. Students wishing to see some of these tiny organisms can collect a humus sample for analysis in the classroom. A simple procedure utilizing an arrangement known as a Tullgren funnel includes the following steps: 1) Gather in a plastic bag a good handful of the uppermost humus and well-distintegrated plant litter. 2) After loosely plugging a large funnel with a bit of cheesecloth, add the humus and support the funnel over a small jar of alcohol or dilute formaldehyde. 3) Place the funnel and contents under a 25 or 40 watt light bulb for about three days and nights or until the humus or litter is dry. A gooseneck desk lamp is effective as a drying unit. Its heat will drive the near-microscopic animals down the funnel and into the collecting jar. 4) With the aid of a magnifying glass or low power microscope examine the organisms that tend to collect on the surface film of the alcohol or formaldehyde. Samples taken from wooded areas will show a high proportion of mites and springtails.

Once students know how to use a Tullgren funnel they can make comparisons of the organic matter found in different areas. For example, students can compare the organic matter found in the top one or two centimeters of an undisturbed A-horizon with the top one or two centimeters of an equal amount of soil taken from an area recently bulldozed. The bulldozer may have removed the topsoil with its organic matter leaving only subsoil with little organic matter exposed at the surface. Or students can take samples from land where erosion may have denuded the topsoil or plowing may have mixed topsoil and subsoil and compare the organisms to those found in an undisturbed A-horizon.

Soil Testing—Acid or Alkaline?

Productive soil not only has good physical structure and texture but offers nutritional support for plants and for bacteria and fungal organisms active within the soil. Minerals—some in large amounts—are essential for plant growth.

Calcium, an element in lime, is required for many processes carried out within plant cells, and magnesium is needed to make chlorophyll. These minerals also help maintain the granulation stability of the soil; together with organic matter, they keep cultivated soils from being hammered by plowing and tillage into a slush with sealed pores.

There is a simple test for lime that students can conduct. Place a few drops of dilute hydrochloric acid (HCl) on a sample. If lime is present, there will be a fizzing from the release of carbon dioxide. Before your class visits a soil profile exposed in the earth, let them conduct the test on some powdered lime or on limestone rock so that they can see the results of a positive test. A student interested in chemistry can investigate what is occurring chemically to cause the fizzing and report his or her findings to the class.

When students view a soil profile, they can test a bit of soil from each horizon for the presence of lime. If the soil tests positively, they can add a drop or two of dilute HCl to rocks in the soil to see whether they are the source of the lime.

Students living in suburban and rural areas can bring into the class samples of soil taken from their lawns and gardens. Urban youngsters can bring samples of soil in which house plants are growing. Is there lime in the samples? Test and find out.

Soil acidity-alkalinity as measured on the pH scale* is an

---

* See Chapter 4, page 100 for discussion of the meaning of pH.

important soil property since the pH requirement of plants varies. Most common farm plants such as alfalfa, barley, cabbage, corn, lettuce, oats, redtop and rye grasses, soybeans, tomatoes, and wheat thrive in slightly acid to neutral soils with a pH of 6–7. Some plants such as lima beans, carrots, potatoes, and strawberries thrive at a pH of 5.5–6.5. The acidity-alkalinity range influences the chemical changes that take place in the soil and determines the availability to plants of some of the nutrients present.

Students can test pH of soil samples brought into the classroom and found at actual soil profile sites by using a soil pH kit. Kits are available from science supply companies such as Carolina Biological or from manufacturers including La Motte Chemical Products Company (Chestertown, Md. 21620) and Edwards Laboratory (P.O. Box 318, Norwalk, Ohio 44857).

Soil Study—Other Minerals

According to *Observation: A Handbook for Teachers*, a Cornell Rural School Leaflet (New York State College of Agriculture, September 1951), a 300-bushel crop of potatoes takes from the soil about 63 pounds of nitrogen, 27 pounds of phosphorus, and 90 pounds of potash (potassium); to restore these minerals, a farmer must add 1,200 pounds of 5-10-5 fertilizer.

Nitrogen, phosphorus, and potassium are absorbed in large amounts by plants and are essential for growth. Although nitrogen makes up about four-fifths of the air, plants do not use it in the gaseous form; the nitrogen must be incorporated into an inorganic compound such as ammonia or a nitrate before plants can use it. Phosphorus and potassium, which generally are not found free in nature, must be available in the soil in the form of compounds.

A good introduction to a study of the minerals that must be present in the soil to maintain productivity is to investigate the composition of different fertilizers. Fertilizers are generally labeled according to the amounts of N (nitrogen), P_2O_5 (phosphate), and K_2O (potash) they contain. A bag labeled "5-10-5" has 5 percent nitrogen, 10 percent phosphate, and 5 percent potash.

HOW TO TRANSLATE A FERTILIZER LABEL

25-5-10 — Fertilizer grade: the proportions of primary nutrients in order of total nitrogen (25%), phosphate (5%) and potash (10%)

MINIMUM GUARANTEED ANALYSIS

TOTAL NITROGEN (N) 25.0%
   8.0% Water Insoluble Nitrogen (From Ureaformaldehyde).
   Other Nitrogen from Urea, Ammonium Sulphate, Diammonium Phosphate.

AVAILABLE PHOSPHORIC ACID
   (P_2O_5) 5.0%
   Derived from Diammonium Phosphate and Super Phosphate

SOLUBLE POTASH (K_2O) 10.0%
   Derived from Muriate of Potash.

Potential Acidity Equivalent to 900 lbs. of Calcium Carbonate Per Ton.

Sears Superfine Lawn Food supplies two kinds of Nitrogen necessary for a good lawn:

   Fast Acting Nitrogen (from Urea) to produce quick growth.

   Slow Acting, even release Nitrogen from Ureaform to sustain growth over extended periods.

Ureaform and Urea act as a team to furnish the Nitrogen needed by lawns.

This means that water insoluble nitrogen, or WIN, accounts for 8% of the bag's total contents. But of the total nitrogen available, what percentage is WIN? Here's a formula for a 40-pound bag:

$$\frac{\text{Poundage of WIN}}{\text{Total nitrogen poundage}} = \frac{8\% \text{ of } 40 \text{ lb.}}{25\% \text{ of } 40 \text{ lb.}}$$

$$= \frac{3.2}{10} \text{ or } 32\%$$

Water soluble nitrogen, though unspecified, probably amounts to 68% of the total nitrogen available (100% less 32%)

If you wanted to neutralize the potential acidifying effect of this product you would add 900 pounds of calcium carbonate for each 2000 pounds of fertilizer. Fine—if you use the stuff by the ton and know that calcium carbonate is a liming material. Translation: Use enough limestone so that its calcium carbonate content equals nine pounds for every 20 pounds of fertilizer.

Water soluble nitrogen

Water insoluble nitrogen, or WIN

With 40 pounds of this fertilizer, nutrient poundage would be, from the top: total nitrogen, 10 pounds (25% of 40); water-insoluble nitrogen, 3.2 pounds (8% of 40); water-soluble nitrogen, 6.8 pounds (10 pounds less 3.2 pounds); phosphate, 2 pounds (5% of 40); potash, 4 pounds (10% of 40)

Copyright 1971 by Consumers Union of United States, Inc., Mount Vernon, N.Y. 10550. Reprinted by permission from *Consumer Reports*, May 1971.

Students living in rural or suburban areas can bring to class descriptive labels cut from fertilizer bags. To guide their investigation, they can follow this question sequence: What nutrients do the fertilizers generally contain? What proportion of nutrients do fertilizers contain? How much fertilizer should be used with different kinds of plants? The nutrient needs of crops differ; alfalfa, corn and tomatoes require high amounts of nutrients, while grasses and pine trees require small amounts. How much fertilizer should be used under different environmental conditions? Plants growing in dryer land will require less fertilizer than plants growing in wetter areas. Is a fertilizer fast-acting? slow-acting?

As a follow-up activity young people living in rural or suburban areas can ask farmers, garden shop proprietors, and green thumb neighborhood hobbyists which fertilizers have proved to be best suited for growing different crops in local fields and gardens. A bus trip to a farm can give city young people a firsthand experience with soil conditions, fertilizer requirements and methods, and other crop-soil relationships.

To gain an even more direct impression of fertilizer nutrient uptakes by plants, both urban and suburban students can spread fertilizer on small plots of grass. Fertilizers high in nitrogen will produce a rapid greening of grass blades. Students can make out plots of several square meters each on a grassy area. On one plot they sprinkle a small handful of commercial fertilizer; on an adjacent plot the amount spread is doubled; on a third plot the amount is halved; a fourth plot is marked off as a control with no fertilizer added. Remind students to water the fertilizer in.

A number of variations of this procedure can be attempted: try equal amounts of fertilizers of different formulae on adjacent plots; fertilize boxes of sod growing in the classroom with different amounts of nutrients; try fast-acting and slow-acting fertilizers on adjacent plots to determine long-term effects, or fertilize with and without adding lime. With most of these experiments, it is probably wise to use a not-too-prominent piece of school lawn; underuse, overuse, or spillage can result in clearly visible burned spots of killed grass.

Still another variation is to test the soil for the presence of nutrients before adding fertilizers and choosing a fertilizer that appears to meet the needs of the soil. Combination soil test kits that include materials, reagents, and directions for making tests for soil pH, nitrogen, phosphorus, and potassium are available from several sources. The La Motte Chemical Products Company sells a soil kit, with twenty tests for each factor for about $11; it includes two booklets containing soil study information. If you are thinking about purchasing a kit, ask for its Educational Products Division catalog. The Sudbury Home Garden Model (twenty tests for each factor) is made by Sudbury Laboratory, Inc. (Sudbury, Mass. 01776). It sells for about $7 and is available from mail order seed companies such as Park Seed Company (P.O. Box 3, Greenwood, S.C. 29647) Some garden supply shops also sell soil test kits.

CHAPTER 6

"THE THREAT OF SUFFOCATION"
Activities with Air Pollution

Today man so arrogantly misuses the atmosphere in his rush toward increased comforts that he is very close to suffocating himself and destroying all the life around him.

That must not happen. We must prevent it. Before the sun is further dimmed, before more flowers wither and die, before more people suffocate in a rank and poisoned world, we must change our ways and learn to live within the laws of nature. We must learn the facts of life.

Air Pollution Primer, National Tuberculosis and Respiratory Disease Association.*

Smoke, fog, smog, haze, and assorted acrid smells assault our noses and lungs so commonly that we sometimes do not realize their pervasiveness and danger unless we inhale a direct blast of bus exhaust, see the stark contrast of clean and foul air masses as an airplane in which we are riding comes into a city for its landing, endure a room heavy with cigarette smoke while suffering a head cold, or see an animal with bones distorted or a plant with leaves burned from chemical poisons carried by the air. These problems are *not* beyond human con-

---
* This familiar organization has changed its name to the American Lung Association.

trol; steps can be taken to lessen what the *Air Pollution Primer* terms the "threat of suffocation."

Students can come to grips with some aspects of air pollution problems by investigating the nature of air, sources of air pollution, effects of air contaminants, and possible solutions. Through such investigations, young people may learn "to live within the laws of nature."

INVESTIGATING AIR

Air Is Really Real?

Primary-grade youngsters may have little conception of the reality of air. Although they inhale the air around them, a glass is "empty" unless it visibly contains a liquid such as water or a solid such as sand. That an "empty" glass is full of gases may at first seem implausible to youngsters who are bound conceptually to the concrete.

A series of simple activities can help beginners develop a conception of air as real. Here are a few that teachers have found useful in the past:

Display three containers—one with water, one with sand, one with air. Students examine each container and write a description of it. Children will tend to describe the third container as empty. Raise the question "Is that container really empty?"

A sign placed above a fish tank of water, a stoppered sink, or a dishpan full of water can ask, "Can you show that the empty glass is really full?" Place several clear tumblers nearby as well as some corks and rubber tubing. As an option during independent study, students individually or in pairs can go to the laboratory center to see if they can devise a way to show that something is in a tumbler. Techniques students can discover for themselves include:

Plunging the mouth of the tumbler into the water tank,

checking the level of the water in the tumbler, and tipping it slightly to let the gas bubble upward out of the tumbler.

Plunging a tumbler in mouth down and placing a long piece of rubber tubing into it. The investigator inhales air from the glass into her or his mouth through the tubing and watches the level of the water rise up into the glass. Students will discover that they need to pinch one end of the tubing until it reaches the tumbler's air cavity to avoid sucking water up into the tube.

Putting a piece of tape on the inside of a tumbler and plunging it mouth down into a tank of water. When the tumbler is pulled out of the water, students test the tape to see if it got wet.

Plunging two tumblers simultaneously into the water, one mouth down so that it does not fill with water, one mouth up so that water fills it. Then the investigator tilts the glass of air so that the air slowly bubbles out and is caught in the other tumbler inverted and held above it.

Schedule a class sharing time so that student investigators can show and explain their discoveries to classmates. Talking about what happens is an important aspect of concept development.

On another day place a glass funnel attached to a long piece of plastic or rubber tubing by the water tank. This time the investigative problem reads, "Plunge the wide end of the funnel into the tank of water. Do you think the water will fill the funnel or stay outside as it did with the tumbler?" Students must record their hypotheses in their science notebooks before actually performing the investigation. Then they see if they can discover why the water does rise up into the funnel. The question becomes, "Where did the air in the funnel go?" Students must prove their explanations (The air went up the tubing into the room) by devising an experiment. This is done simply by holding a finger over the end of the tubing while lowering the connected funnel into the tank of water; the water will not enter the funnel for the air cannot escape.

Punch a small hole into a side of a plastic bottle that has a tight screw-top lid; a plastic ammonia or shampoo bottle will serve nicely. Print on a card the following directions:

Put a finger tightly over the hole you will find on the side of the plastic container. Fill the container with water, holding it in the sink. Have your partner screw on the top. Hypothesize whether the water will come out when you take your finger away.

Now try the experiment. Hold the bottle straight up and over the sink and remove your finger from the hole.

If you find that the water does come out, think about why this happens and write your explanation in your notebook.

If you find that the water does not come out, think about why this happens and write your explanation in your notebook. See if you can figure out several different ways to make the water come out. Be prepared to explain why your ways work.

Post the card above a science table where the necessary equipment has been laid out. Partners can go to the table to investigate the problem during independent study times.

After most of the class have had the opportunity to investigate the effects of air pressure, schedule a group talk session so that students can share their findings and explanations. Mount a blank chart on an easel so that one student can serve as scribe and record general findings and conclusions. The explanation to work toward is that air on the outside of the bottle held the water in and that air can exert pressure.

Set up a series of test investigations based on the principles

"The Threat of Suffocation" 141

already developed. Make available at the science center a glass or clear plastic tube about 15 centimeters long, soda straws, medicine droppers, a bottle with a narrow neck, clay, a funnel, a cork, and a container of water. On large index cards (call them science cards with the students), print the following directions:

Lower the tube into the water. What happens to the water? How far up the tube does it rise? Record your observations in your science notebook.

Now lower the tube into the water, but this time hold a finger over the end of the tube. First, hypothesize what will happen. How far up the tube will the water rise? Why? Record your preliminary hypothesis in your notebook. Try the experiment to verify your prediction. Record your observations.

Experiment with a soda straw. Lower the straw partly into the water. Inhale and exhale with your mouth on the straw while a partner watches the level of water in the tube. Record your discoveries about how a straw works.

Draw some water up into the straw. Keep your tongue over the top, and remove the straw from the water. Do you think the water will stay in the tube? Why? Do you think the water will stay in when you remove your tongue? Try the experiment to verify your prediction.

Experiment with the medicine dropper. Then formulate an explanation to tell how the dropper works. Draw a diagram to help explain.

> Put the end of the funnel into the top of the narrow-necked bottle and mold a ring of clay around the neck so that the connection between bottle and funnel is airtight. Now rapidly pour some water into the funnel. Record what happens. Explain why this happens.
>
> Next very slowly pour water into the funnel. Record what happens. Explain why this happens.

> Use any combination of materials (funnel, cork, water, bottle, and so forth) to demonstrate in your own way the presence of air around us and/or the pressure air exerts.

Gather students who have completed an activity card together in groups to talk about their discoveries and share their written reports. Help students to verbalize the major concepts—that air is all around and that it exerts pressure.

What Is Air?

Air—the stuff we breathe—is a mixture of gases; the precise amount of each component varies from place to place as conditions change. Generally, however, the mixture of gases by volume is about:

    78.088% nitrogen
    20.949% oxygen
      .93% argon
      .03% carbon dioxide
      .003% other gases (helium, methane, neon)

In addition, air has some water vapor in it; the amount varies from practically none to 4 percent by volume.

Students can graph the percentages by volume of the gases present in air as they begin to build a concept of what air is. Ask some students to express the data in bar graphs, while others express the data on line graphs.

There are a number of activities through which upper-elementary young people can develop understanding of what air is all about. An easy beginning is to discover the presence of water vapor in the air. Three or four students working in teams fill a dry metal can (a soup can does nicely) with ice cubes and stand the can on a piece of paper. Gradually the sides of the can will become wet, and the paper becomes moist as water drips down on it. A scribe systematically records on large charting paper what the group did and what they observed. They attempt to explain where the water came from. Could it have come through the can? over the sides of the can? from the air? When groups have completed their explorations, the group reporter shares with the class the group's investigative data compiled on the chart.

Because the amount of water vapor in the air may vary considerably from day to day, it would be wise to involve students in the water vapor activity first on a humid day. The activity can be repeated on several occasions so that comparisons can be made.

The presence of oxygen in the air can be demonstrated very simply. Students begin by accepting the fact that burning uses oxygen. Attach a short candle to a small glass plate with melted wax. Lower the candle and plate into water no deeper than one quarter the height of the candle. Light the candle and then cover it with a thick, clear glass tumbler; the level of the water will gradually rise as the burning candle takes oxygen from the air. The candle flame will finally be extinguished when there is not enough oxygen to support combustion. Students can infer that the oxygen is diminishing as water rises and takes its place.

Remove the tumbler quickly from the water and cover it to

save its contents. Then add limewater to the contents to demonstrate the presence of carbon dioxide in the residual air. Limewater in the presence of carbon dioxide turns from clear to milky. When limewater is added to the tumbler in which the candle was burned, it will become milky.*

This activity works rather well as a teacher or student demonstration; the teacher can first help a group of students carry out the activity and the group can then demonstrate and explain to the class.

INVESTIGATING CONTAMINANTS IN THE AIR

How We Contaminate the Air

The *Air Pollution Primer* distributed by the American Lung Association categorizes air-contaminating activities into three general classes: 1) attrition, 2) combustion, and 3) vaporization.

To investigate how attrition—the wearing or grinding down of materials by friction—contaminates the air, students rub a number of materials against each other. A simple kit to study attrition might contain: a file, a piece of sandpaper, a nail, steel wool, the heel on a shoe, a piece of wood, a piece of plastic, grains of rice, a leaf, pine needles, cement, a rock, a scrap of aluminum foil, and a noodle. Students can put together their own discovery kits and test to see whether one object, such as sandpaper, will loosen particles from another material on which it is rubbed back and forth. By keeping a record of the degree of difficulty in loosening particles by abrasion, they can go on to hypothesize which materials are worn down most easily and where in nature particles are added to the air by attrition.

---

* Limewater can be purchased from a drugstore or a chemical supply house, ready to use. However, powdered lime can be added to water and the liquid filtered to produce limewater inexpensively.

A recording guide such as the one shown below may help students in compiling data:

```
                      ATTRITION ANALYSIS
                                                    SHOE
                  KEY   SANDPAPER   NAIL   FILE   HEEL   ETC.
   Wood
   Foil
   Rock
   Pine Needle
KEY
 X    Abrades with Difficulty
 XX   Abrades with Moderate Effort
 XXX  Abrades Easily
```

Because ferrous metal to metal contact can produce dustlike iron particles, a magnet can be used to investigate the results of attrition. Students chalk off a square about 35 x 35 centimeters on the pavement near or on a road. With slow, careful movements, they sweep the entire marked-off area with a strong magnet held a few millimeters from the ground. If the magnet is placed in a thin plastic bag of the kind used to wrap sandwiches, the metal fibers and chips picked up can be collected by turning the bag inside out before removing the magnet. The fibers and chips can be analyzed under a microscope or magnifying glass, and students can compare the kinds and amounts of iron particles picked up at different locations and on different occasions. They might even investigate the classroom floor; iron particles dropped by floor polishing equipment or carried in on students' shoes may be discovered.

A series of teacher demonstrations can show how combustion contaminates the air. With tongs hold a sheet of metal above a burning candle. A sooty film will appear on the metal surface. Try holding the metal surface about a meter above the flame. Does a film appear at this distance from the burning wax? Slowly move the metal collecting surface closer to the flame to obtain a greater amount of air contaminants. If a magnesium ribbon is available, you might carefully light the end of a small strip while grasping the other end with long tongs. Students will see white particles being emitted into the air as a result of this combustion. (Note: *Do not hold the burning strip with the fingers. Hold the strip above a fireproof surface.*)

Do the same demonstration using a match, paper, a cigarette, and steel wool as the burning source of contaminants. Does the absence of a collected film imply that no contaminants are released?

A slightly more complex demonstration shows that some air contaminants are actually invisible. Place some wood chips on a wire gauze laid on a tripod and ignite the chips with a propane torch, a Bunsen burner, or even an alcohol burner. Construct a collecting bottle in which there is limewater to test for the presence of carbon dioxide (see the diagram). Above the burning wood chips, hold a funnel that is connected to the collecting bottle with rubber tubing. A student can slowly operate a vacuum pump to draw air from the collecting bottle and allow the gases produced through the combustion to bubble into the limewater. As combustion proceeds, the limewater turns milky indicating that carbon dioxide, a colorless gas, is being given off as a product of the burning.

This experiment can be repeated using other combustible materials such as paper, matches, Bunsen burner gas flame, or charcoal. You may even have students try steel wool, which, because it does not contain carbon, will not emit carbon dioxide. Young people may be able to hypothesize what kinds of materials will emit carbon dioxide into the air when burned, and what kinds will not.

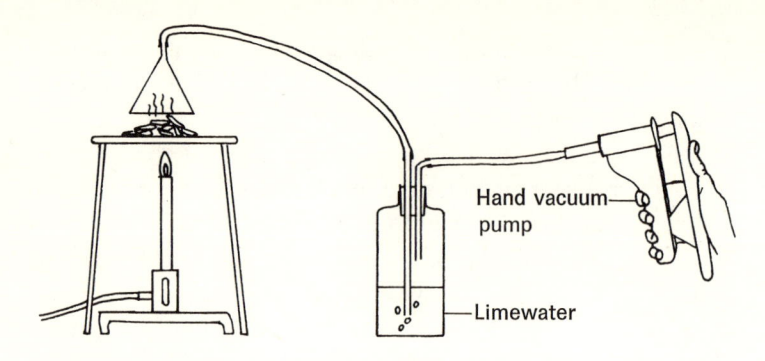

TESTING BURNING FUEL FOR CARBON DIOXIDE PRODUCTION

Vaporization, or the conversion of liquids to gases, can be demonstrated rather simply. Pour perfume into a low-rimmed vessel such as the lid from a margarine tub or a petri dish. Students stand at one foot intervals from it. Rapidly fan across the dish's surface. As students smell the vaporizing perfume, they indicate by raising their hands.

Older students can attempt a more systematic investigation of the distances materials travel in the air. They hold litmus paper moistened in distilled water at varying distances from ammonia escaping from an open bottle. The litmus will turn blue at distances where there is vaporized ammonia. (Note: *Remember that ammonia should not be inhaled.*)

Sometimes a vaporized chemical interacts in the air with a second vaporized chemical to form a pollutant that may be more dangerous or offensive than either of the original chemicals. Such a combination can be observed by bringing an open bottle of hydrochloric acid very near to an open bottle of ammonia (ammonium hydroxide). A dense cloud of "smoke" will form at the point where the vapors from the two bottles come into contact. This cloud is ammonium chloride, a new compound that has formed. In this case, of course, ammonium chloride is less toxic than either the hydrochloric acid or the ammonium hydroxide from which it was formed.

In addition to attrition, combustion, and vaporization, people contaminate by directly introducing particles into the air. One such source of contamination is aerosol spray liquid particles introduced under pressure into the air. Room fresheners, deodorants, hair sprays, insecticides, paint sprays, and oven sprays are among the most widely used home aerosols. Students can begin to investigate this source of air contamination by studying the components in aerosol containers as indicated on the labels and warnings on the cans. Put several varieties of household aerosols in the science center for students to take notes about during independent activity times.

Seeing Particles in the Air

How clean is the air in a classroom? How accurate is the nose in judging air cleanliness? A simple demonstration can show students that the nose cannot always judge the extent of particulate material in the air.

On a sunny day darken the classroom, allowing only a small shaft of sunlight to penetrate. Students can experiment with viewing the light beam from several angles; they will quickly discover that dust and fine dirt particles are clearly visible to observers looking up the beam of light but are almost invisible to observers viewing the beam from the side. Students will be amazed by the multitude of particles present in the air around them.

The impression can be heightened if a cigarette is briefly burned and held in the light beam; the particulate material in the air will increase. Seeing the actual cigarette smoke may make young people more aware of the pollution smokers add to the air.

Vacuuming the Air

Since foot traffic brings most of the large particulate matter into the air of buildings, students can try sampling the air of the school for particles they track in on their shoes.

An old vacuum cleaner can serve as the sampling instrument, or the school custodian may have an industrial vacuum cleaner that students can use for their investigation. Perhaps students can even enlist the custodian to become part of the experimental team.

To trap the dirt particles when air is drawn through the cleaner, fasten a piece of filter paper or cloth over the hose opening. Students may discover that they need first to fasten a piece of ordinary window screening over the hose opening as a backing so that the filter paper or cloth will not be sucked into the hose. They may also have to experiment to find what kind of cloth is coarse enough to let air through but fine enough to trap particles.

The vacuum cleaner must run for a long time; professional high volume air samplers run twenty-four hours to collect a sample. The time necessary can be gauged by watching for the filter to become darkened. This time must be recorded if samples taken in several areas are to be compared.

Some possible investigations are 1) sampling a busy entrance hallway and a quiet, little-trafficked place and comparing the samples collected after equal periods of time; 2) sampling the air outside the school and comparing it to a sample collected at a busy entrance hallway; or 3) sampling the air in the cafeteria, the gymnasium, and library, making comparisons, and hypothesizing the reasons for differences.

Sampling Particulates with Sticky Traps

Sticky traps can be used to gather large particles that pollute the air. Because there are numerous ways of making sticky traps, students can work in teams, each employing a different technique for sampling particulates in the air.

The Vaseline Trap: Students smear glass or clear plastic

microscope slides with a very thin coat of Vaseline and place the slides where they wish to sample the air.* Traps are left in place from one day to a week. When samples are collected, students examine them with a low power microscope or hand lens to see the kinds of particles collected. They can estimate the amounts collected by comparing their specimens to the photographic standard for particles per square inch that can be obtained from Technical Associates for Industry, Inc. (P. O. Box 116, Park Ridge, N. J. 07656).

---PARTICLES PER SQUARE INCH--- OVER 20 MICRONS
CINCINNATI VISUAL STANDARDS A-3
CITY OF CINCINNATI AIR POLLUTION CONTROL AND HEATING INSPECTION

1000 2000 3500 5000 10,000 20,000 30,000 45,000 60,000

Used by permission. The above reproduced one-half size should not be used as a reference standard. The correct official standard may be obtained from TAFI, Inc., P.O. Box 116, Park Ridge, N.J. 07656.

If students place several sticky traps in each location, collection periods can be compared; slides left out for one day can be compared to those left out two days, four days, or a week. Similarly, by placing traps in various locations, the effects of differing heights from the ground, differing traffic patterns, and differing air circulation patterns can be investigated.

This technique can be used not only to sample indoor loca-

---

* Silicone stopcock grease used by chemists works even better than Vaseline. Students may want to experiment with a variety of greasy materials to see which ones make the best particle traps.

"The Threat of Suffocation" 151

tions such as window sills, bookcase surfaces, and chalk troughs but also to sample outdoor sites. To catch particles swept by wind, place the traps in a vertical position. Students can nail clamp-type clothes pins to trees, poles, or other vertical structures and then clip the glass slides in place.

The Punch-Hole Trap: Using a paper punch make a neat row of holes through an index card and then fasten a strip of transparent sticky tape over the holes. A series of such cards can be exposed in various places for different lengths of time. Students can examine the particles caught in the punch-hole traps with a magnifying glass. Holes on one card can be visually compared to holes on another card to obtain data on the numbers and kinds of particles collected at each location. Again students can compare their specimens to the photographic standard obtained from Technical Associates.

To obtain a series of specimens varying in exposure time, students can punch a series of holes in a piece of thick cardboard and fasten strips of transparent sticky tape over the holes on both sides of the cardboard so that the sticky sides of the tape face each other. At the beginning of the experimental period, one tape is pulled back from one hole, exposing the sticky side of the second tape. The next day the tape is pulled back from another hole, exposing more tape, and so forth day by day. At the end of the period students, using microscopes or magnifying glasses, can compare holes on the tape that have been exposed for successively greater time periods.

The Cardboard Slide Trap: Students can also make collecting traps from the cardboard frames of discarded 35mm color slides. Strips of transparent sticky tape or contact paper can be fastened across the frames and exposed in different locations for various time periods. Students can project their samples with a slide projector, allowing the entire class to inspect their data. By the way, if students want to preserve their samples, they can spray the trapped particulate material with clear lacquer.

The Directional Bottle Trap: The Cincinnati Division of Air Pollution Control has used adhesive-coated paper wrapped around circular glass jars for collecting windblown particles. Wind speed, precipitation, and the diameter of the collecting jars are important factors affecting particulate collection.

Students can experiment with fastening sticky paper strips—stickyside outward—on baby food, olive, and mayonnaise jars and can compare the collecting efficiency of small and large jars. Each strip on the jars should be marked to indicate its placement, e.g., north, south, west, east, or streetside, fieldside. To gain better exposure and to hold jars firmly in place, jar lids can be nailed upside down to boards or posts before the collecting jars are screwed on.

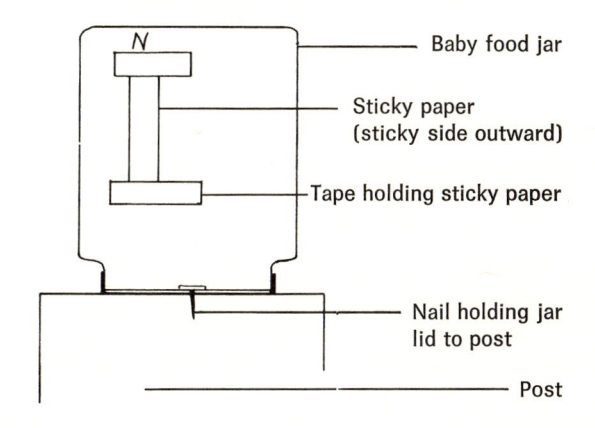

DIRECTIONAL COLLECTOR FOR WINDBLOWN PARTICLES

Students can compare the strips by peeling the sticky papers from the jar and taping them to a piece of transparent projection paper. The data can then be projected and enlarged for viewing, and students can compare the particle buildup in different locations and in different directions.

"The Threat of Suffocation"

If students want to try this investigation on a grand scale with a high grade sticky surface, Fasson Products (250 Chester St., Painesville, Ohio 44077) can furnish a minimum order carton of a hundred 7 x 22-inch sheets of self-adhesive Pli-a-Print R-135 for about $19.

Sampling with Dustfall Catchers

Because wind blows particulates about in the air, young people have seen dust and dirt from the air settle by gravity onto window sills and shiny surfaces of cars. By placing large dustfall catchers on the schoolyard or in their own neighborhoods, students can determine what and how much is settling from the air in specific locations during given time periods. This activity is most effective in urban areas where black, sooty materials tend to settle out of the air.

Simple dustfall catchers can be made from large jars or pails. Widemouth gallon plastic jars of the kind in which salad dressing is packaged may be obtained for the asking from restaurants and diners. Some ice cream is sold in large plastic pails that also are suitable catchers.

Student investigators firmly anchor the empty, clean dustfall catchers in open areas so that normal air movement is not blocked by an adjacent wall or roof overhang. Jars should also be placed at least several meters above the ground to keep surface dust from blowing in and to prevent them from being knocked over or carried off by animals and children.

Jars can be anchored by wedging them between two scrap blocks of wood nailed to a board or fastened to a board by wire hooked over the lip and nailed to a board under the jar. An alternate method is to bore small holes in the neck of the jar and hang it on a washline with wire.

The catchers should be placed at several diverse locations for a minimum of one month and then brought inside for analysis. Students record the dates at beginning and end of the sampling time and write descriptions of the experimental locations. At the end of the sampling period, they examine the jars visually and pick out any leaves, paper, or debris that are not particles settled out of the air. If there is water in the catchers, they must next transfer the collected sample into a small pyrex beaker or a small enameled saucepan and take the following steps:

1) Add small amounts of hot water to the catcher, scrape with a rubber or plastic spatula if necessary to free the solids, and pour the contents of the catcher into the beaker. 2) Boil the sample *almost* to dryness on a hot plate. 3) Remove the beaker or saucepan, letting the remainder dry without added heat. If the container is not removed, the sample may burn and stick to the bottom. 4) Feel the material with the fingers. Is it sandy? gritty? tarry? 5) Describe in words how much was collected in a month.

If a balance that can weigh to one tenth of a gram or less is available, very capable students can weigh the small dried sample. Many single- or double-pan beam balances commonly have sliding weights that weigh to a tenth of a gram. A spring scale will probably be inadequate. First weigh the container with the dried sample. Then clean and thoroughly dry the container and reweigh. The difference between the two weighings is the weight of the sample.

Samples commonly weigh less than a gram, but that is only the amount that fell through the collecting jar neck. Students can estimate or calculate to get some idea of how much material settles from the air onto a square meter or an entire neighborhood.

Checking Dirt in City Air

A simpler way to check the accumulation of dirt on surfaces, especially in cities where incinerators, furnaces, jet planes, buses, and cars are continuously throwing soot into the air is to staple a large piece of white construction paper to a heavy piece of corrugated cardboard to make a sturdy collecting surface. Position the collecting surface on an open, flat surface, holding down the corners with bricks or rocks. Place the lids from margarine tubs in several rows across the surface, holding each lid down with a weight. Remove a lid every two days, jotting on the paper the date the lid was removed. Observe the dirt buildup in the circular areas and compare those uncovered early in the sequence, and thus exposed longer, to areas uncovered later. Differences may be evident when collections of weekdays are compared to those made on weekends or business holidays. The sampler sheet must be brought inside during wet weather.

Sampling Snow

If your school is located in an area that receives winter snowfall, sampling the particulate matter intercepted in snow is a project a team of upper-grade students can try. Professors Moran and Morgan of the University of Wisconsin at Green Bay describe procedures and applications in an article titled "Snow Sampling: A Student Project for Determination of Urban Airborne Particulate Distribution" in the November 1973 *Science Teacher*. They suggest collecting samples of the upper one inch of undisturbed snow cover and melting and filtering the samples through no. 2 filter paper. Students dry and weigh the filter papers before and after the filtering operation. The before weight is substracted from the after weight to get the actual weight of the material.

The meltwater caught running through a funnel is measured in a graduated cylinder to find out its volume in milliliters. To compare the amount of particulate material collected at different sampling spots, students divide the weight of the sample caught on the filter paper by the volume of the snow meltwater. The result is the weight of particulate material for each milliliter volume of meltwater collected. It is this weight that students should compare as they sample snow collected at different locations and on different occasions.

Our Community

Major sources of air contaminants are automobiles, airplanes, industrial plants such as cement factories, steel mills, and chemical processors, electric power generators, heating systems of homes, offices, factories, and schools, burning trash, and use of pesticides (see James Marshall's *The Air We Live In*, Coward, McCann, 1968, pp. 38–49). According to *The Complete Ecology Fact Book* edited by Philip Nobile (Doubleday/Anchor, 1972, p. 196) the percent by weight of air pollution emissions in 1968 was:

Transportation	42%
Fuel combustion in stationary sources, e.g., electric generators, heating systems	21%
Industrial processes	14%
Forest fires	8%
Solid waste disposal	5%
Miscellaneous	10%

Students who are learning about visual ways to present data can plot on a graph the percent by weight of air pollution emissions, experimenting with horizontal and vertical bar graphs, line graphs, circle graphs, and pictographs to determine which type is most effective for presenting these data.

Once students have become aware of the categories of air pollution emissions, they can use their understanding to survey pollution sources in their own community. They can list

AIR POLLUTION SOURCES

Miscellaneous ⋈ ⋈

Solid Waste Disposal 🔥

Forest Fires 🌲 🌲

Industrial Processes 🏭 🏭 🏭

Stationary Combustion 🏢 🏢 🏢 🏢 ᴇ

Transportation 🚗 🚗 🚗 🚗 🚗 🚗 🚗 🚗

One Picture Represents 5 percent by Weight of Air Pollution Emissions.

---

specific sources—the local cookie manufacturing plant, a metals mill, a refinery, pesticides used in farming, or a major highway—and compile a complete listing of local polluters. Studying these data, investigators can estimate the percentages of emissions that fall into the categories already identified, and their estimates can be graphed and compared to national averages.

Air Pollution Mapping

In urban centers, outdoor air pollution is generally concentrated at certain points and along certain lines because of the location of fixed pollution sources, traffic congestion patterns, and weather conditions, including wind direction and speed.

Using their noses to gauge the extent of air pollution and possibly the degree of eye irritation as a secondary gauge, teams of students walk through different streets of their community with notebooks in hand identifying areas of high air pollution. Findings may include:

☐ Heavier amounts of noxious fumes at traffic stop intersections than at other intersections because of acceleration and deceleration of vehicles.

☐ More pollution smells on major arteries than on side streets.
☐ More noxious fumes on truck and bus routes than on residential streets.
☐ More eye irritating fumes by diesel train lines than by electric train or subway lines.
☐ More vehicle exhaust smells on department store basement and ground floors than on higher floors.
☐ More fumes or dust clouds near a factory emitting materials than in streets farther away from the factory.
☐ Greater pollution downwind from a factory than on the upwind side.

Before observation teams go out, discussion can help investigators formulate ways to observe and build the vocabulary necessary to record descriptions of things observed. Be careful, however, that the discussion does not prejudice the observers; recordings should reflect actual observations, not hypotheses.

Upon returning to the classroom with their data, upper-grade students can make a large grid map showing labeled streets and key landmarks. Using the erasers on the ends of pencils with a stamp pad students put many ink blobs in the streets and other locations of high pollution and fewer, more scattered ink blobs in areas of low pollution concentration.

Lower-grade youngsters can draw a large street map on big sheets of brown wrapping paper. Small squares of different colors can be cut from construction paper and pasted on the map. Black squares may indicate high pollution areas; green squares may show low pollution areas. Factories and other buildings made from cardboard boxes may be placed along the streets to show urban concentration; plastic cars and trucks can be added to show traffic patterns.

In rural areas this activity can be adapted to indicate pollution caused by pesticide spraying. Students can make field maps to show fields where pesticide pollution is high or low.

Smog Weather Watch

Smog usually is worst when the air is stable with little or no movement to carry away or disperse the pollutants being generated. Because of this relationship, students can correlate smog development with weather information.

They can use visibility of landmarks as a subjective measure of smog buildup. Road maps can indicate approximate distances to chosen landmarks so that some are a half kilometer away and others are farther away where tall buildings or mountain peaks project on the skyline. Each morning and afternoon a class weather observer records which of the chosen landmarks are clearly visible or are less visible. She or he also records wind speed and cloud cover. Students make up a graded wind speed list by noting wind effects on objects. Their list might include one of the following:

Tree leaves motionless, flag hanging limp—no wind
Leaves lightly stirring—slight breeze
Papers lifted from the ground—gentle breeze
Papers moved along the ground—moderate wind
Tree branches and/or metal signs
  swinging—strong wind
Tree branches swinging and breaking
  to the ground—very heavy wind

Cloud cover can be expressed in terms of the amount of sky covered. No clouds would be recorded as 0/10, sky half covered as 5/10, and completely overcast as 10/10.

After landmark visibility and weather data are collected over a period of weeks, students analyze their findings to see if smog periods can be correlated with windy or calm weather, with clear days or cloudy.* If charted for longer periods, students

---

* Remember that if the humidity is very high, visibility reduction may be due to plain fog—not smog.

can find if smog occurs during warm or cold periods, fall, winter, or spring. They can check newspaper and TV weather reports to see how their findings compare to national weather observations.

By the way, this is an ideal context in which to introduce students to Lewis Carroll's conception of portmanteau words. Both *smog* and *smaze* are portmanteau words formed by combining two existing words to produce a word that carries some of the meaning of both of the original words:

$$\text{smog} = \text{smoke} + \text{fog or a smoke-filled fog}$$
$$\text{smaze} = \text{smoke} + \text{haze or a smoky haze.}$$

Upper-grade students can go on to invent their own portmanteau pollution words such as *suffocill* from suffocate and kill, or *stoke* from stink and choke.

West Coast as well as other cities periodically experience smog alerts when the quantity of pollutants in the air becomes excessive; warning flags are displayed to tell citizens that it may be dangerous to exercise to any great extent. Using information supplied by TV and newspaper forecasters and their own estimates of air pollution levels, students can monitor and announce local air pollution conditions. They make flags that

Low pollution level 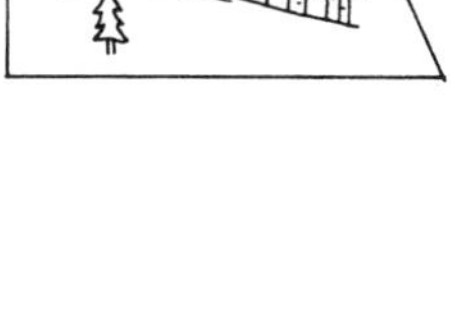   Moderate pollution level 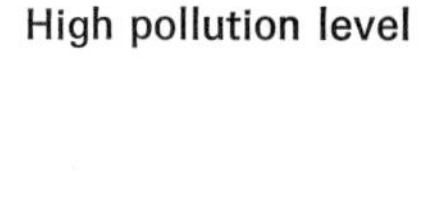   High pollution level

stand for the air pollution levels and fly the appropriate one based on their estimates and news reports.

Winds and Pollution

Wind carries many air contaminants away from the offending source. Because molecules of hot air are further apart from each other than molecules of cold air, a volume of hot air is lighter than an equal volume of cold air. As a result, hot air tends to be displaced upward, carrying pollutants with it.

There are several ways to investigate the rising of hot air currents. Upper-elementary students can make a spiral paper "snake" cut out of a circle of paper and suspend the "snake" over a heat source such as a hot plate. The snake will whirl, propelled by the upward draft of hot rising air. Similarly, a plastic pinwheel will rotate when held above a hot radiator, again demonstrating the upward movement of hot air currents.

Students in teams can fasten a balloon over the lip of a soda bottle and heat the bottle by plunging it into hot water or standing it on a radiator. If a pyrex flask is available, the balloon can be fastened around the neck of the flask and the base heated gently with a Bunsen or propane burner to obtain faster and more spectacular results. In either case, before heating the bottle, students should hypothesize what the effects of the heat will be; then they record pictorially and verbally what actually happens.

That hot air tends to move upward can be shown by measuring the temperature at different elevations in the classroom, first a few millimeters from the floor and thereafter at meter intervals. Data are shown visually on a line graph in which height in meters is shown on the vertical or y-axis while temperature in degrees Celcius or Fahrenheit is plotted on the horizontal or x-axis.

To make a visual and subjective determination of a wind pattern, an observer can study smoke as it leaves a chimney, especially industrial smokestacks. When updrafts are strong, smoke may disperse rapidly; under other conditions, smoke may hover around the stack, move sideways appearing as a cone- or fan-shaped layer in the sky, or even be carried earthward.

An art or photography related student project is to draw or photograph smokestack patterns on different days and to hypothesize wind direction based on patterns of smoke dispersal observed. Other influencing factors the observer must consider include the height of the stack, the velocity at which smoke is emitted from it, ground temperature, downdrafts from buildings and nearby hills, and so forth.

Two excellent color filmstrips titled *Air Pollution,* Parts I and II distributed by Diana Wyllie Ltd. (3 Park Rd., Baker Street, London NW 1) for about $8 per strip, illustrate smokestack patterns and weather phenomena that influence these patterns. A booklet describing each frame accompanies the strips. The series provides a helpful resource for junior and senior high school young people involved in a pollution study.

An investigator studying smokestack patterns as an individual project may wish to go on to judge the density of smoke emitted using a handy comparative grid called the Ringelmann Smoke Chart, a chart that serves as a standard. For example, the Los Angeles County Air Pollution Control District limits to three minutes in any hour the discharge of any air contaminant from a single source that is as dark or darker than that designated as number 2 on the Ringelmann chart. A student researcher can check sources in the local community that fail to meet this Los Angeles standard. A plastic Ringelmann smoke indicator, such as that reproduced here, can be purchased from Enviresearch Corporation(7 Dalamar St., Gaithersburg, Md. 20760) or from Ward's Natural Science Establishment for $2. A similar cardboard comparison card is available from McGraw-Hill Publishing Company. The instructions for using the card are:

1. Look through clear windows at arm's length to sight pollution source and match it to a corresponding shade of gray.
2. Read black smoke density and record as Ringelmann number 0, 1, 2, 3, 4, 5 (top scale).
3. Read all other colored emissions in opacities and record as percent (bottom scale).
4. Opacity readings are related to Ringelmann numbers as shown. (Caution: Light source should be behind the observer. *Do Not Look Through Card At Sun or Any Light Source! Damage To Eyes May Result If This Precaution Is Not Observed.*)
5. Readings should be made at right angles to wind direction at any convenient distance which provides clear view of stack and background.

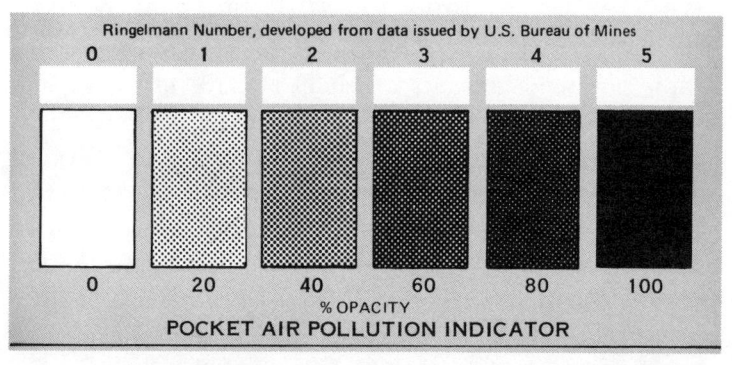

Ringelmann Number, developed from data issued by U.S. Bureau of Mines
0 1 2 3 4 5
0 20 40 60 80 100
% OPACITY
POCKET AIR POLLUTION INDICATOR

Used by permission of Enviresearch Corp., 7 Dalamar St., Gaithersburg, Md. 20760. Over 25,000 now in use.

An Inversion Lid on the PAN

Normally the temperature of air decreases as elevation increases; at an elevation of one kilometer, one would expect the temperature to be about 6½ degree Celcius (or Centigrade) cooler than at sea level and at an elevation of two kilometers, one would expect the temperature of the atmosphere to be 13 degrees Celcius cooler. The reason relates to the physical laws governing the behavior of gases.

When talking about this phenomenon, students may remember that if they have climbed a mountain, the air became cooler; they may recall seeing pictures of mountain tops covered with snow even in the summer when surrounding lower elevations are considerably warmer. To trigger a discussion of this complex topic, you might bring in pictures of Mt. Kilimanjaro and ask, "Why is there snow on Mt. Kilimanjaro if the

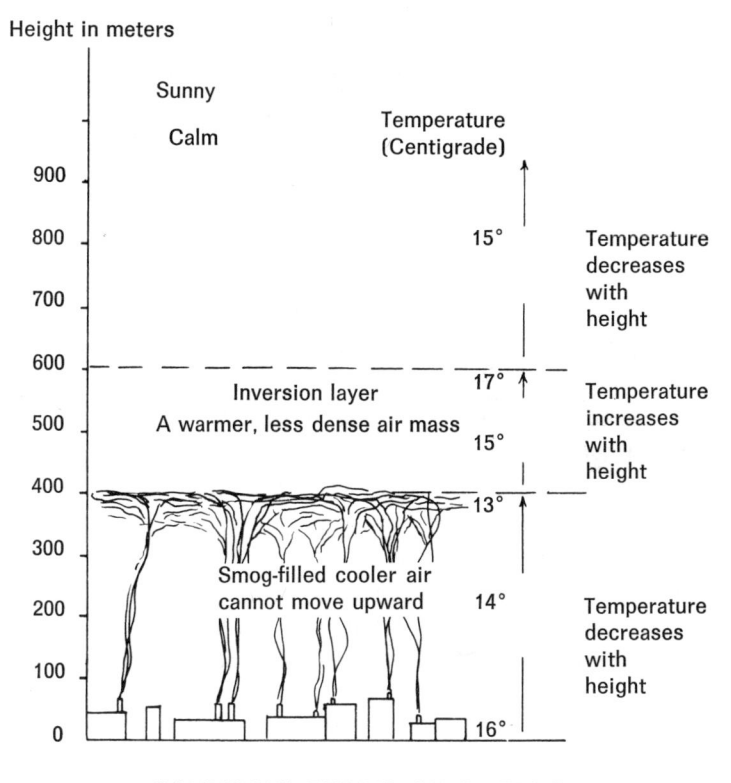

Height in meters

Sunny
Calm

Temperature (Centigrade)

Height	Temperature	
900		
800	15°	Temperature decreases with height
700		
600	— — — — — — — — — 17°	Temperature increases with height
500	Inversion layer — A warmer, less dense air mass — 15°	
400	— — — — — — — — — 13°	
300	Smog-filled cooler air cannot move upward	
200	14°	Temperature decreases with height
100		
0	16°	

CONDITIONS DURING AN INVERSION

mountain is located just below the equator?" The answer, of course, is that Kilimanjaro towers 5,963 meters above sea level.

Sometimes, however, the surface of the earth and the air adjacent cool rapidly. When this happens, a layer of relatively warmer air is sandwiched between the rapidly cooled bottom air layer and the colder upper layers. This warm air layer acts like a lid and stops the cooler and heavier bottom air layer from moving upward and prevents pollutants emitted into that cooler layer from being lifted away. Meteorologists call this situation a surface inversion. Inversions become pollution traps, holding air with a high level of pollutants close to the ground.

Another kind of inversion occurs aloft rather than close to the ground. Called subsidence inversions, the higher ones are caused when a mass of high pressure air moves into an area and warms as it settles down (subsides), forming a lid over the cooler air below. These warmer high pressure masses can approach close to the surface and remain for days. The clear skies and gentle winds that accompany a high pressure mass can cause a surface inversion to form below.

Weather forecasters often talk about inversions, because inversions can lead to dangerous health conditions. Unburned gases from gasoline engines contribute toxic nitrogen oxide and ozone-generating smog pollutants to the air. To make matters worse, the ultraviolet energy of the sun triggers the combination of ozone and these unburned engine gases to form compounds called PAN (peroxyacetyl nitrate). PAN compounds are eye irritants and cause heavy tearing. They also severely damage plant leaves during periods of inversion.

In *For Pollution Fighters Only* (McGraw-Hill, 1971, pp. 83–4) Margaret Hyde describes a demonstration that can explain how an inversion operates. She suggests that a student select two jars equal in size and place one in a refrigerator or pan of ice cold water and the other in a room of average temperature. After the refrigerated air has had time to cool, the student removes the jar from the refrigerator and inserts smoke by hold-

Downtown Los Angeles under clear skies

Downtown Los Angeles with inversion beginning to build up

Downtown Los Angeles under heavy smog

—Air Pollution Control District, County of Los Angeles

ing burning punk, paper, a cigarette, or rope into it. Immediately she or he inverts the warm air jar over the cold air jar. The smoke stays in the bottom jar because the cold air there is heavier than the hot air above. The investigator should describe what happens and should see if she or he can explain the phenomenon and what the effects on air pollution levels might be.

If you work with older young people of junior high age or above, you may wish to carry on the more sophisticated investigation described in the State of California's Department of Public Health brochure, *Experiments for the Science Classroom Based on Air Pollution Problems,* revised edition, 1962 (2151 Berkeley Way, Berkeley, Calif. 94704). This brochure shows an inversion box, which is reproduced with permission.

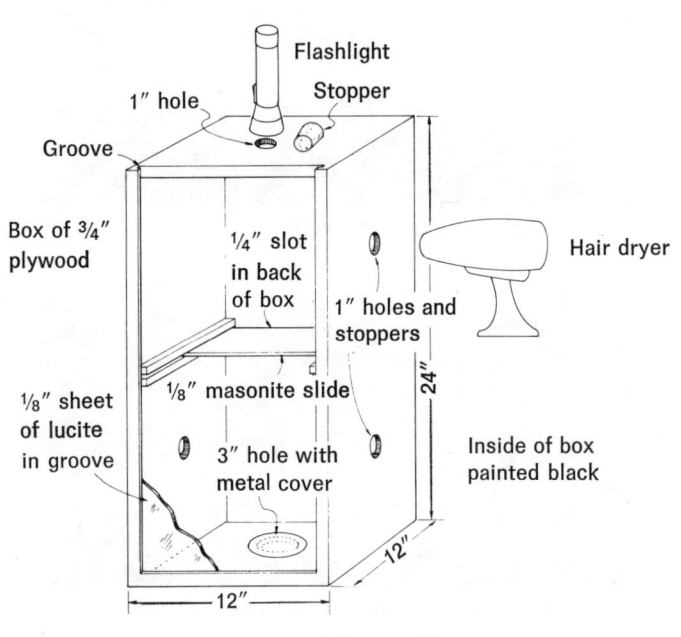

INVERSION BOX

In inversion box investigations, students introduce warm air into the upper half of the box using a source such as a hair dryer. At this point, the masonite or wood slide partition, which separates the top and bottom compartments, is in place; the top hole in the upper compartment is open to provide a vent for the heating air and to test the temperature of the vented air. While the air in the upper compartment is being heated, the holes in the bottom compartment are securely capped.

When the air coming out of the top vent hole feels rather warm, students stopper all the holes in the upper chamber and pull through the rear slot the masonite partition separating top and bottom compartments. Then they unstopper one of the holes in the side of the bottom compartment and introduce smoke using a burning paper, cigarette, or punk. With lights extinguished and a flashlight beamed downward through the hole at the top, students can observe the smoke layer and hypothesize why the smoke stays in the bottom compartment and does not enter the top compartment.

Students next move the box so that the three-inch hole covered with metal at the bottom of the lower compartment hangs over the table surface; they heat the metal cover with a candle, carefully watching what happens in the above chambers. They will see that the inversion is broken and that the polluted layer now moves upward. This corresponds to the sun's heating the ground, after which the warmed surface air can break through the inversion layer above.

INVESTIGATING EFFECTS OF AIR CONTAMINANTS

Danger to Man!

The warning carried by cigarette advertisements, "The Surgeon General has determined that cigarette smoking is dangerous to your health," indicates a relationship that exists between inhaled air contaminants and human health. Students can study some aspects of this relationship in the following series of activities:

They can visit a local hospital or invite an X-ray technician to talk to the class. As part of the visit or talk, students examine X-rays of lungs. Hospitals may have pictures of the lungs of people who lived most of their lives in cities to compare to people who have lived most of their lives in the country; they may have pictures of the lungs of heavy, moderate, and nonsmokers.

Investigate chronic respiratory diseases such as bronchial asthma, chronic bronchitis, pulmonary emphysema, and lung cancer as well as lesser conditions related to air contaminants such as eye tearing, headache, and dizziness. The American Lung Association's (formerly the National Tuberculosis and Respiratory Disease Association) *Air Pollution Primer* is a valuable source of information that students in grades five and above can use for research. The Association also has available a series called "The Facts," which are four- to eight-page leaflets on various respiratory diseases that are useful references for junior high school students.

Investigate the contaminants produced by burning cigarettes. A filtering device for collecting contaminants is easy to construct. Hold the mouths of two small identical funnels together, place a piece of ordinary filter paper and a piece of screening for support between them and squeeze modeling clay around the connection to make a tight fit. Insert a cigarette into the stem of one funnel, place a piece of rubber tubing over the stem of the other funnel and connect it to a vacuum pump. Light the cigarette and operate the pump slowly to draw air through the makeshift filtering device.

Students can repeat the experiment using a filter-tipped cigarette and varying the time of exposure. They can do research to discover the nature of pollutants emitted by cigarettes. (Note: *Students should handle the tarry filter with tweezers, for although smokers inhale these products, they are dangerous.*)

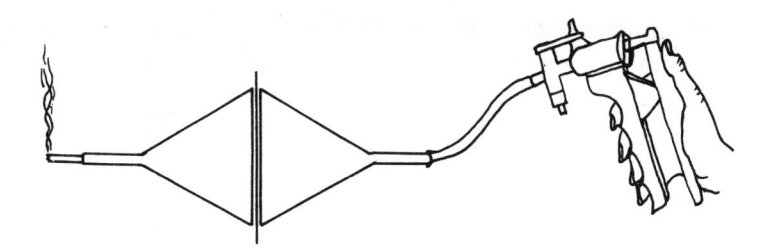

SIMPLE FILTERING APPARATUS

In the second edition of Morholt, Brandwein, and Joseph's *A Sourcebook for the Biological Sciences* (Harcourt Brace and Jovanovich, 1966, p. 288), a simple method for estimating human lung capacity is described. Students fill a gallon bottle about four-fifths full of water. They insert a two-hole stopper, one hole of which contains a short piece of glass tubing that does not extend into the water in the bottle; the other hole contains a longer piece of tubing that extends well down into the water.

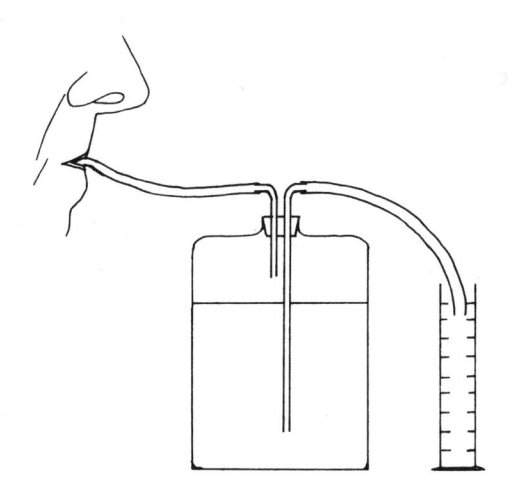

LUNG CAPACITY ESTIMATOR

The authors suggest that students seal the bottle top with paraffin but modeling clay will also serve to make the connections airtight. Students connect pieces of rubber tubing to the ends of both tubes extending out through the two-hole stopper. One tube leads into a graduated cylinder; students take turns exhaling into the other tube, covering the mouthpiece with paper toweling for sanitary reasons. As students exhale the air from their lungs, the air will go into the collecting bottle and displace water, which can be measured in the graduated cylinder.

Students who have built this device to test lung capacity may set up their equipment on open school night and investigate parents' lung capacities. If parents also complete a questionnaire on which they indicate smoking habits, occupation, exercise habits, exposure to noxious particles and gases, and present and former residences, students can later consider if there is a relationship between any of these factors and lung capacity.

And Plants Too!

The leaves of tomato plants become blotchy white after exposure to air pollutants; endives grown in polluted air are stunted in growth. Spinach, tobacco, grapes, and a host of other plants are damaged by even a slight amount of ozone in the air. Pollutants that cause such damages include chlorine, ethylene, hydrogen fluoride, nitrogen dioxide, ozone, PAN, and sulfur dioxide; they enter a plant through either leaves or roots.

To study the effects of air contaminants on plant growth, students systematically compare plants growing adjacent to a heavily trafficked highway to plants living at specified distances from the same road. They examine plants next to the road, 3 meters away, 5 meters away, and 10 meters away looking for differences in the number of plants, growth development, color —specifically blotchiness, bronzing, silvering, and burned edges —and number of flowers or pine cones (reproductive parts).

Students who live near industrial plants can make a similar study, comparing vegetation close to the factory to vegetation at measured distances from it.

A helpful resource for students who wish to investigate plant damage caused by air contaminants is a stunning full-color brochure *Air Pollution Injury to Vegetation*, published by the U. S. Department of Health, Education, and Welfare in 1970, and distributed by the Government Printing Office (Washington, D. C. 20402) for $1.25. The brochure contains color photos of vegetation damaged by pollutants and diagrams of leaf and root structures showing how pollutants enter; the pictures project nicely with an opaque projector. The brochure also warns against mistaking insect damage for pollution damage.

Because some mineral deficiencies as well as insect damage can be mistaken for pollutant damage by the untrained eye, the county agricultural agent may be a helpful resource person for students attempting to distinguish pollutant from other forms of damage. Maybe the agent will talk informally with students, especially in rural areas where the economy is highly dependent on agricultural production.

A different approach to the study of air pollution injury to vegetation relies on a method outlined by Arnold Darlington in *Ecology of Refuse Tips* (London: Heinemann Educational Books Ltd., 1969, pp. 53–6) to test for the presence of contaminants on leaves and pine needles. Darlington collected samples of contaminants by pulling pine needles and privet hedge leaves through pieces of "folded paper gripped by a small bulldog clip padded with cork." Narrow smears of visible pollutants corresponding to the lines of the needles or midribs of the leaves were produced.

Student investigators can do the same by simply wiping pine needles or leaves with folded pieces of tissue grasped between their thumbs and forefingers; white toilet tissue works well because equal-sized squares are already marked off by the perforations. Samples should be taken from needles or leaves that are equidistant from the ground and that face the same direction. Samples can be taken close to a polluting source such as a highway or factory and at progressively greater distances away, both upwind and downwind from a source, and before and after a rain. Back in the classroom students can photograph and/or write descriptions of their samples.

Still another activity is to grow seedlings and then expose the plants to contaminants. Tomatoes, sunflowers, and beans are rather easy to raise from seed. Once seedlings are five to ten centimeters tall, build a frame with coat hanger wire over the flat in which the plants are growing, cover the frame with a plastic sheet, attaching it to the wire with clip clothes pins, and through a hole in the plastic introduce a pollutant found around the house. Aerosols (hair spray, room freshener, or oven spray) can be sprayed into the plastic hood for a specified time. Keep the hood over the plants for about an hour or so after spraying; then remove until the next exposure period.

Students must remember to keep a control flat; the control receives the same treatment as the experimental flat (including hood on and off) except for the introduction of the pollutant. Students may wish to take photographs periodically so that they have a record of the development of the control and experimental plants. In any case, a systematic written record should be kept of size of plants and lushness of growth as well as of color characteristics.

Nature's Threats

Volcanoes: Students usually picture flows of molten lava when they think of a volcanic eruption, but remind them that the gases given off are also considerable. While water vapor is the major component of the gases—60 to 90 percent—noxious and deadly gases include carbon dioxide, chlorine, hydrochloric acid, hydrogen fluoride, hydrogen sulfide, and sulfur dioxide.

It is estimated that after the eruption of Mt. Katmai in 1912, the Valley of Ten Thousand Smokes in Alaska emitted over a million tons of hydrochloric acid.

To demonstrate, put a few drops of hydrochloric acid on pieces of wood, paper, and limestone. Students can draw parallels between the corrosive action observed and the effect on lung tissue of breathing hydrochloric acid vapors even in small concentrations. Gaseous hydrogen chloride becomes hydrochloric acid when it hits moisture in air or lungs. (Note: *Elementary school youngsters should probably not work with hydrochloric acid. Do this one yourself.*)

Students will enjoy making a volcanic cone from plaster of Paris so that they can stage an "eruption." Ignited ammonium dichromate crystals placed in a small depression in the top of the "volcano" will burn brightly and simulate a cinder-type lava flow as the resulting green powder cascades down the sides of the cone. A five centimeter strip of magnesium ribbon ignited as a fuse will add smoke and greater brilliance to the eruption. (Note: *Again, do the igniting yourself.*)

Forest Fires: Forest, brush, and grass fires are obvious sources of air pollutants. Many fires are set by lightning, but student awareness of man's role in fire prevention can be increased. Students can compile their own caution lists, which might include making sure campfires are out before leaving a campsite, keeping matches away from young children, not smoking or making fires when woods are dry, not throwing glowing cigarettes from car windows, and heeding forest fire alerts. Younger students may enjoy drawing Smokey the Bear cartoons to drive home the message of fire prevention.

Pine "Smells": When we walk through a quiet pine woods, we may feel that we are close to nature and associate the smell of pine with fresh air. Unfortunately, too much pine "smell" may damage lung tissue. Dr. George Smith at the University of Nevada's Desert Research Institute suggests that terpene, the hydrocarbon that gives pine trees their aroma, may cause emphysema and asthma. Cattle emphysema has been a common problem for ranchers in the Southwest. The cause of this condition, called the "grunts" because of the cattle's labored breathing, has never been completely clear. The grunts occur in the fall when cattle are brought down into the valleys where terpene concentrations are heavy. In summer, grazing in the mountains where there is little terpene in the air, the cattle never seem to be troubled. Dr. Smith and his associates are experimenting with cattle breathing terpene, but more definitive work is needed to link asthma and emphysema in man to exposure to terpenes.

Students living in the cattle ranch areas near the Sierra Nevadas or in the Blue Ridge Mountains of the Southeast can investigate the terpene problem by asking veterinarians if they have examined cattle with the grunts, inquiring health agencies if they have had cases of emphysema that possibly could be related to terpenes, or asking ranchers if their cattle have shown the seasonal pattern of grunts described above.

Pollen: Sticky traps such as Vaseline-coated glass slides or transparent sticky tapes catch not only particulate matter such as dust and bits of plant fiber but also airborne pollen grains, especially during the flowering periods of many trees, grasses, and obnoxious plants including ragweed.

EXAMPLES OF POLLEN GRAIN SHAPES

Pollen grains of different plant groups have their own shapes and markings. The illustration depicts several kinds. Student investigators can identify pollen particulates with a microscope that can magnify 100 times or more. They can compare traps exposed during different seasons, and students who suffer from allergic reactions can attempt to correlate periods when their symptoms are greater with periods when traps exhibited larger "catches" of pollen.

Atomic Radiation: Scientists express concern about the increasing use of materials that emit radiation. There is, of course, natural radiation produced by the normal breakdown of radioactive materials. In addition to this natural radiation, today radiation is being emitted in the production of power by atomic reactors, in bomb tests, and in medical diagnosis and treatment. Since radiation is known to trigger some forms of tissue damage as well as genetic changes, this contaminant of the air does pose a threat to life.

Although young people cannot experiment with the effects of radioactive isotopes on plant or animal growth because of the dangers involved, they can become aware of the presence of radiation in the air around them. Borrow a Geiger counter from the high school science department; if the high school does not own one, a technician from a local college or hospital might come to demonstrate one. The counter can pick up the radiation emitted by the luminescent dial of a watch and test for differences in background radiation.

Damage to Materials

A mix of organic gases, produced by incomplete combustion of gasoline and oxides of nitrogen and emitted by industries and utilities, will react to form ozone, a compound that damages plants, destroys rubber, and fades dyed fabrics. Students living in an industrial or heavy trafficked area can expose rubber and dyed cloth to the atmosphere to investigate possible ozone damage.

Place pieces of cloth and short lengths of rubber hose in a shelter that blocks direct sunlight but allows air to pass through. You or the custodian can place the shelter on the school roof or in another undisturbed location, and the sample can be examined after one-, two-, and three-month periods. It is best to fasten the cloth to a simple cardboard or plastic frame and to stretch the rubber a bit before nailing it to a board or hanging it from a hook with a weight on its end. The presence of ozone will be shown by the fading of the cloth and the cracking of the rubber. Duplicate control samples should be kept wrapped and inside for comparison purposes.

If there is room in the shelter, students can add a paint test by making triplicate samples of various paints on scrap strips of wood. One sample goes in the shelter out of direct sunlight, one is exposed nearby to the atmosphere and sunlight, and the third is kept covered inside as a control. Soiling effects of soot or oil film will show more quickly than will the wearing qualities of the paints. Paint companies expose test panels for several years to determine long-term effects.

Women's nylon stockings are susceptible to deterioration from a number of air pollutants and can be used to demonstrate the presence of contaminants as well as show an adverse economic effect in daily living.

Pieces of nylon must be exposed so that air can freely move through the threads. Stretch pieces of nylon stocking over short sections of cardboard cylinders taken from rolls of toilet tissue or paper toweling and fasten them with elastic bands. Suspend these roll mounts horizontally by tacking them to a wood surface or the side of a post. Or mount nylon pieces in Polaroid #633 3¼ x 4-inch slide mounts to permit freer movement of air through the nylon. Suspend slide mounts by wedging each into a slot cut in a support or by suspending them across two bricks. Schedule exposures of thirty, sixty, and ninety days. Keep another sample in an envelope as a control for later comparison. At the end of the exposure periods, with a magnifying glass examine the nylon for breaks. This activity is particularly appropriate as an individual or team project.

Silver that is exposed to hydrogen sulfide in the air rather quickly loses its characteristic shine and becomes a dull brownish black. Students can systematically investigate this reaction by wrapping in plastic bags several silver spoons that have been freshly cleaned and polished. Each week remove a spoon from the plastic bag and place it with the experimental group that is exposed to air. At the end of each week, compare the newly exposed spoon to those exposed on previous weeks to check for progressive discoloration.

As an example of the extreme effects of hydrogen sulfide on silver, suggest that students eat a hard boiled egg with a silver spoon. After only a short period of contact with the egg, the spoon becomes tarnished.

Documenting Air Pollution

Encourage students to compile photographic essays to document and add interest to an air pollution study. Investigators can take photos or cut pictures from newspapers and magazines showing the following:

☐ Views of distant landmarks on clear and smoggy days.
☐ Panoromas of a smog area taken from a high geographic point or tall building to show the depth of the smog or the upper "lid" of the foul air masses.
☐ Pollution sources such as a burning dump, poorly tuned diesel vehicle, belching foundry stack, smoky grass fire, or dust-spewing alfalfa drier.
☐ Effects of pollution on people and things, e.g., people covering their noses with handkerchiefs or wiping their irritated eyes, chemically eaten marble and granite monuments, blackened building façades, soot-covered window sills.

Students who are skilled at sketching may draw pictures of many of the same sorts of occurrences.

Design a bulletin board titled Insult Board and subtitled What We Do to Our Lungs to depict visually the effects of air pollutants on human lungs. Divide the board area into about fifteen rectangles. Students search newspapers and discarded magazines for pictures or draw their own sketches and mount them in one of the rectangles and label them with appropriate "insults."

POLLEN	PIPE, CIGAR AND CIGARETTE SMOKE	COAL DUST INHALED BY MINERS (BLACK LUNG DISEASE)
FUNGUS SPORES	AUTO EXHAUST	METALLIC DUST AND GRINDINGS INHALED BY METAL WORKERS
VIRUSES	INCINERATOR SOOT AND FUMES	ASBESTOS FIBERS
BACTERIA	FUMES FROM REFINERIES AND SMELTERS	URANIUM DUST INHALED BY MINERS
TALC POWDER	AEROSOLS	COTTON DUST INHALED BY TEXTILE WORKERS

CLEANING UP THE AIR

Clean Air?

What do people think about air pollution? What are people willing to sacrifice to achieve clean air? Dr. Robert Rankin of West Virginia University investigated public attitudes and reactions in his area. Students can conduct similar investigations. They may want to find out if people:

Are aware of and concerned about air pollution.
Believe they are personally being affected.
Think air pollution affects health.
Agree that there is or is not a problem in their community.
Are willing to drive their automobiles less and ride public transportation, lower home temperatures, use electricity in their homes more sparingly, accept less outside lighting, or accept less refrigerated fruits and vegetables to cut down on air pollution.
Think that air pollution is confined to an area immediately around the source.
Know of any efforts by federal or regional agencies to deal with air pollution in their area.

They could also ask what people associate with air pollution (dust? chemicals? odors?). Students can devise their own specific questions to include in a survey of public opinion, and their results can be compiled into a research report.

Knowing that the automobile is one of the worst polluters, young people may wish to find what can be done to reduce the contaminants produced by cars and trucks. Some paths of action are: 1) Check driving habits of family and friends: Is the car allowed to idle unnecessarily? Does the driver make jackrabbit starts and stops? Is the car engine kept in tune? 2) Ask a mechanic to explain pollution control devices on the car and the way they function. 3) Visit a motor vehicle inspection station or repair garage to see the operation of exhaust emission test equipment. 4) Conduct a library study to find how electric, Wankel rotary, and steam autos operate and to find advantages and disadvantages of these engines. The antique car buff will, of course, be interested in reporting on the Stanley steamer.

Don't Burn the Leaves

Until recently leaf burning was common practice in suburbia. Every fall neighborhoods smelled of thick, acrid smoke as fires burned along the streets or in trash burners. Greater public consciousness as well as laws banning leaf burning has reduced the problem, but the question of what to do with the leaves remains. Some communities collect leaves for a community compost heap; residents can later obtain the organic matter decomposed from the leaves for their gardens.

Building a compost pile either in the schoolyard or at home serves several purposes. It provides a continuing source of biological specimens, becomes a point of interest to be checked at intervals during the year, and eliminates the leaf and grass disposition problem.

There are many successful ways to build a compost pile. Here is one. Start with a 20-centimeter pile of leaves, weeds, grass clippings, or a mix of these vegetable materials. Wet the layer, and then add 5 centimeters or so of animal matter—manure, sewage sludge, or some other good nitrogen source, such as wilted green grass clippings, which are high in nitrogen. Next sprinkle on a little soil or old compost to provide a source of microorganisms. Repeat the layers as materials accumulate. Make the compost pile about a meter wide and high so that it does not dry out but stays damp (not soggy). If the compost starts to smell, it needs air. Drive some holes in it with a stake or turn it over.

Microbial action in the center of a compost pile can raise the temperature to 60° Celcius (140° F). This is an advantage since weed seeds are killed at such temperatures. Students can test their compost piles for the extent of microbial action by measuring the temperature. They can determine the "readiness" of their compost by checking the extent of decay; it is ready when it is a fairly uniform crumbly mass in which the original leaves and stems are no longer identifiable.

Students may wish to start a planting project with their homemade organic fertilizer and experiment with not using chemical pesticides. An excellent resource for those with little gardening experience is Rita Reemer's *Teaching Organic Gardening* (Emmaus, Pa.: Rodale Press, 1973); the text provides information and activities on how to plan, start, and maintain an organic garden and includes sections on composting, insect control, and even organic cooking. *The Organic Classroom* by Thomas Fegely and Bud Souders (Rodale Press, 1973) is a helpful reference for students in grades four and above. It too provides activities on composting and pest control for the beginning organic gardener.

A Clean Stack

To keep particulates such as dust and fly ash out of the air, manufacturing and power plants must attach air pollution control equipment to their furnaces and other processing units. Industries that need these very elaborate devices include papermaking, steel manufacture, lime and Portland cement kilns, and electric utilities. At Bethlehem Steel's Lackawanna (New York) plant, control devices collect 190 tons of dust daily, which would otherwise be discharged into the air!

Visits may be arranged to local utilities or manufacturing plants to see air pollution control equipment, or students can find in library references what the devices do. Most devices are variations of baghouse dust collectors, wet scrubbers, or electrostatic precipitators. In a sense, baghouse collectors can be compared to household vacuum cleaners, wet scrubbers to flushing dirt off a house with a watering hose, and electrostatic precipitators to the static electricity effect one gets when a comb is used to pick up pieces of paper or shreds of styrofoam plastic after running the comb through dry hair or rubbing it on wool cloth. Bethlehem Steel Corporation (Bethlehem, Pa. 18016) offers an attractive booklet *Keep It Clean,* which explains pollution controls with color photos and diagrams.

Individual Versus Community Rights

Does an individual have the right to emit air contaminants that may affect the health and well-being of others in the community? Does the community have the right to interfere with an individual's pleasure and/or ability to turn a profit? Some general questions students can informally consider or debate in specific contexts are: Should the government regulate the amount and kinds of contaminants that cars, buses, trains, and planes can legally emit? Should the government restrict use of cars in areas where air pollution is intense? Should local communities establish leaf and trash burning ordinances? Should the government regulate the use of pesticides sprayed into the air, especially by crop dusting planes? Should the government legislate against people smoking in public conveyances and buildings—airplanes, trains, restaurants, or public meeting rooms? Should the government regulate the amount and kinds of contaminants an industrial plant can discard into the air? Should the government help to pay for pollution control equipment needed by industry? Should the government tax the worth of air control equipment as it does the rest of the industrial plant? Should the government ban the sale of items that may damage the health of individuals if research evidence is not conclusive?

Students in four-person teams can select a controversial issue, research it, and present the pros and cons to the class in a "Listen to Both Sides of the Issue" round-table discussion. If the session is taped, a summarizing team can later listen to it, summarize in writing the points made, and prepare a mimeographed handout for general distribution.

A Job for You

Because the causes, effects, and controls of air pollution touch so many facets of daily living, a study of air pollution problems introduces students to a range of occupations. Possible career options open to students include:

In the health sciences
    Physicians specializing in diseases of the chest
    Medical laboratory technicians
    X-ray technicians
    Health physicists
    Public health nurses

In engineering
    Mechanical, chemical, industrial, sanitary, and traffic engineers

In other applied sciences
    Meteorologists
    Biologists specializing in pollution studies
    Agricultural agents

In other technical areas
    Instrument repair technicians
    Sandblast cleaning contractors
    Auto mechanics specializing in pollution control devices
    Pollution control inspectors

Students in grades five through eight can research an occupation, finding out what a worker does, the importance of the job, the educational requirements, the location of jobs, and so forth. Girls should be aware that these highly skilled occupations are as available to them as to boys.

A Song for You

Song of the Smoggy Stars by Osmond Molarsky (Walck, 1972) is a picture book for younger students that tells the story of Thaddeus, a boy who decides to do something to make the stars sparkle brightly in his city. Armed with his guitar, Thaddeus takes up the fight against air pollution. He writes the *Song of the Smoggy Stars*, tapes his song, and plays it to convince people to bundle rather than burn their newspapers. As with other Thaddeus stories by Molarsky, both words and music of the song are included in the book.

Song of the Smoggy Stars can be read aloud to youngsters in sixth grade and below. It can motivate students to conduct their own clean-up-the-air campaign as well as to write their own air pollution ballads. Like Thaddeus, they can tape renditions of their songs and share them with others.

With older young people, Hamlet's words or other poetry can serve as a trigger to motivate students to write air pollution laments:

> this most excellent canopy, the air
> look you, this brave o'erhanging firmament, this
> majestical roof fretted with golden fire, why, it appears
> no other thing to me
> than a foul and pestilent congregation of vapours.
> *Hamlet*, Act II, Scene 2

Start this activity as a total class experience in which students begin by brainstorming ideas about the foulness of air; ideas are recorded on the board as they are projected. Later writing groups select ideas, order them, and add others to form simple ballads or laments. The ballads or laments can be printed on the recycled paper students learn to make in Chapter 3; the finished products can be bound together to form a booklet with a title created by the students.

CHAPTER 7

ENERGY CHALLENGES
Activities with Energy Problems

. . . children growing up in the next several years are likely to have fewer material comforts than their parents enjoyed during their childhoods.

An overstatement? Quite possibly.

An understatement? Also quite possibly.

Whichever, the age of continually rising expectations . . . has been halted, at least temporarily, by the energy crisis. Increasingly people are beginning to realize that more of everything is not always attainable simply with technology or money.

William D. Smith, *New York Times,* January 6, 1974

Drivers have found long lines at the gas pumps. Workers have faced job losses because of cutbacks in purchases of large cars, mobile homes, decorative outdoor lighting, and in travel. People have been turning down their thermostats and turning off their lights. Such conditions testify to the fact that the energy shortage is no longer just a prediction of scientists who are acutely aware of energy relationships. The energy crunch is upon us!

There are no simple answers to the shortage, for energy issues are highly interrelated with problems of air pollution, soil and land destruction, and disposal of materials used to produce energy. The shortage has geopolitical, ethnic, and social overtones as well. And underlying the energy shortage problems are basic scientific considerations:

☐ The earth's supply of stored energy is limited.
☐ As stored energy is consumed for fuel or is converted into electrical or chemical forms to drive motors and machines, losses are constantly occurring; much energy is dissipated as heat and escapes from the earth.
☐ Energy cannot be recycled; we must constantly obtain more from inside the earth or from the sun.

As young people in classrooms study these complex interrelationships of the energy situation, they must become involved in geopolitical, social, environmental, and scientific studies. They must begin to understand that they as individuals have a role to play in possible solutions. Perhaps they may come to feel as did Marcy Enoch, an eighth-grade student from Carthage, Tennessee, who wrote, "I would rather have a warm home than a ride in a car . . . . Bike riding is also very healthy for you as well as your country."

Or as Cherri Blair of Dallas wrote, "Sometimes when my parents forget to turn down the thermostat, I try to remind them about it. Sometimes my smaller brothers and sisters forget to turn off the television, lights, and radio, and I try to remind them."

Or as Tony Marshall of Dearborn, Michigan, explained, "I used to watch TV all night long, and I stopped it. I am cutting down a little bit at a time. I still watch it, but not as much." (*New York Times*, January 27, 1974)

INVESTIGATING ENERGY SOURCES

Energy from Coal

Coal originated in swamps or bogs where lack of oxygen prevented layers of dead, compacted vegetation from completely decaying. Most coal mined today was formed several hundred million years ago from the luxuriant vegetation that existed during the geologic periods known as the Pennsylvanian, Permian, and Cretaceous.

Student interest in the Age of Reptiles—the periods between the Permian and Cretaceous—and in the dinosaurs that roamed the earth at that time has always been high. Capitalize on that interest to involve students in investigations into areas such as life in the coal-forming swamps, how peat becomes coal, or the eras and periods of the geologic time scale.

Students living near a bog or swamp can conduct a firsthand investigation. It can include an analysis of the degree of decay of organic matter, peat formation, oxygen content of the water, and pH to find conditions possibly comparable to beginnings of coal formation.

Several films describe the processes involved in the formation of the earth's coal reserves. McGraw-Hill distributes a fifteen minute color film *Coal: A Source of Energy* that tells how coal was formed millions of years ago and how it is being mined today. This film works well with youngsters in grades four and up; check your local film library for this or a comparable presentation.

You can raise a number of questions with your students about coal used as fuel: What happens to coal when it is burned? What is the difference between anthracite and bituminous coal? between bituminous coal and peat? Does all coal produce the same amount of sulfur when burned? the same amount of heat? What are the problems associated with using coal as a fuel? How does coal compare in cost to other fossil fuels such as oil and gas? How is coal transported from mine to consumption site?

To find answers to these questions, students can visit a coal yard to obtain samples of anthracite and bituminous coal and to interview the manager, search encyclopedias for tables that compare the heat value of anthracite and bituminous coal, and write seeking information from the National Coal Association (Coal Building, 1130 Seventeenth St., N.W., Washington, D.C. 20036). In coal mining areas, they can direct their inquiries to the local mining company.

In some areas, public utilities faced with shortages of petroleum are retooling their plants so that coal can fuel their furnaces. Invite an engineer from a local public utility to speak to your class about problems related to the use of coal as fuel, about the percentage of the utility's operation fueled by coal, and about the utility's attempts to use other energy sources and to protect the environment. You may find that the utility has a slide-and-talk presentation available for schools. For example, Idaho Power, which draws heavily upon water power, sends out a speaker to junior high schools to talk about the scope of the energy crisis, electricity and its relationship to the environment, and the company's operations.

Many geological experts believe that all the major coal fields have been discovered. Encyclopedias generally contain world maps showing the locations of the major coal reserves. Make a transparency of such a map and project it with an overhead projector. Raise the questions: What regions of the world are well supplied with coal? What regions are bereft?

Energy experts estimate that coal is being used at a rate that will deplete the reserves in 100 to 400 years. Some chemists believe that a large portion of the coal reserves should not be used as fuel but should be conserved as a rich source of valuable organic compounds that can be converted into plastics, dyes, and coal tar chemicals. A simple project for younger children is to investigate the products that can be obtained from coal and to cut pictures from magazines to make wall charts showing the wide range of items that are derived from coal.

For older students arrange a demonstration to show that coal is composed of organic fractions. Coal, when highly heated without oxygen present, will separate into its components—coal gas, coal tar, and coke. Coke is carbon and mineral ash and is produced when volatile gas and tar are driven off. This process is called destructive distillation. To demonstrate destructive distillation, follow these steps:

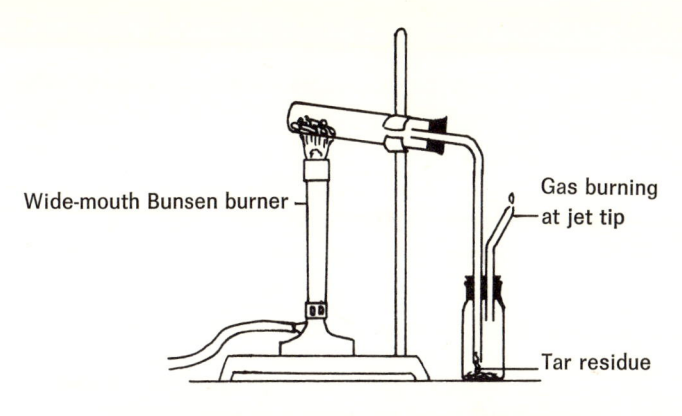

DESTRUCTIVE DISTILLATION APPARATUS

1. Half fill a large pyrex test tube with ground bituminous (soft) coal. Don't use anthracite because it produces only a small percentage of volatile matter whereas bituminous yields over 25 percent.

2. Set up the apparatus as depicted in the diagram.

3. Heat the coal with a Fisher Bunsen burner or a propane torch and observe the tarry liquid that begins to collect in the bottle.

4. Ignite the coal gas as it comes from the glass jet tip. Note that at first only air is driven off and that constant reigniting may be necessary when the coal gas is finally emitted because of the water vapor content.

5. Test for the presence of ammonia by hanging a piece of moistened Hydrion pH paper or red litmus paper down the neck of the bottle before inserting the two-hole stopper. Change of pH paper to alkaline range or red litmus to blue will show base (alkali) present—the ammonia.

6. Remove the coke remaining in the test tube and burn it in a direct flame as a fuel.

Energy Challenges

Energy from Oil and Gas

Experimenting with distillation and/or combustion of petroleum and volatile petroleum products is not advisable in an elementary or junior high school. Materials are dangerously flammable, and the fumes of many petroleum fractions are noxious and explosive. These investigations are best performed in a chemistry laboratory under expert supervision with adequate safeguards. Intermediate and junior high pupils, however, can be involved in the following activities:

Map Studies: Student researchers analyze maps to identify working and potential oil and gas fields, to find routes of existing and projected pipelines, and to identify high consumption areas. Researchers categorize nations in terms of their consumption and production, e.g., nations that have large oil resources and are exporters of oil and/or gas, nations that have limited reserves, and nations that have no reserves and must import.

Students can make map transparencies that show pipelines, reserves, ports, and population centers. If all transparency maps are made to the same scale, several can be superimposed, e.g., rail lines and coal reserves, gas and oil fields, or pipelines and oil fields. In this way interrelationships can be noted. Or students can make a large-scale bulletin board world map on which they plot oil, gas, and coal reserves, pipelines, population centers, and ports. Analysis of data plotted can produce generalizations about energy relationships.

Model Making: Pupils can construct models of oil and gas related phenomena such as Drake's original oil rig in Titusville, Pennsylvania, a cross section of rock strata in which oil is trapped, or the continental shelf showing where offshore drilling rigs are probing for oil and gas. The American Petroleum Institute (1801 K St., N.W., Washington D.C. 20006) as well as individual oil companies will supply diagrams of rock strata and drilling rigs that can serve as guides for model makers. A contour map of a portion of the continental shelf can guide students constructing offshore models.

A working demonstration model of a steam-driven turbine is another possibility. Boil water using a Bunsen burner or propane

Mount wheel on a nail and insert nail into a short stick of wood.

Make cuts with metal snips toward the center of a metal coffee can lid. Twist and flute the cut wedges to make the turbine wheel.

DEMONSTRATION MODEL OF A STREAM-DRIVEN TURBINE

torch flame as an energy source. The water is contained in a flask corked with a one-hole stopper through which extends a piece of bent glass tube with a jet tip. Steam coming through the tubing is directed onto blades, which whirl and simulate turbine movement. The demonstration shows how the energy stored in the gas is converted to mechanical energy. (Note: *Put the stopper in gently and heat flask slowly.*)

By the way, you may be able to locate a functioning model of an old steam engine to demonstrate the same principle. Water in a boiler is heated with a flame to produce steam that drives a piston and flywheel. The demonstration also shows the dissipation of energy. The boiler walls lose heat to the air, and

Energy Challenges

the friction of moving parts generates more waste heat.

Statistical Studies: Many data are available from a variety of sources that detail the comparative consumption of gas, coal, oil, and other fuels. For instance, some brokerage houses supply customers with a breakdown of the energy sources tapped by specific public utilities as shown in the following table:

PERCENTAGE USE OF

COMPANY	NUCLEAR	COAL	GAS	OIL	HYDRO-ELECTRIC
Allegheny Power		99			1
Arizona Public Service		68	30	2	
Florida Power and Light			31	69	
Northern States Power	23	46	21	4	6
Texas Utilities		6	93	1	
Puget Sound Power and Light					100

One brokerage house listing contains a breakdown for eighty-eight companies. In analyzing these data, students can categorize companies by geographical regions and compute for each region the percentage of dependency on oil, gas, coal, nuclear, and water resources for the production of electricity. They can go on to hypothesize why there is such a vast difference in the energy sources utilized by public utilities. Help students identify determining factors such as closeness to ports, closeness to wells and mines, availability of water power sources, state regulations limiting emission of sulfur in high population areas, and cost. Check brokerage houses in your community for these or similar data.

Check also the public utility servicing your region. The annual report of the company contains detailed analyses of profits and losses related to energy problems and a detailed breakdown of fuel sources that the utility taps to produce power. Pollution control equipment being installed may also be reported. You will find many tables and graphs, as well as maps, that students can analyze.

Graphs and tables comparing production of fossil fuels by countries of the world are becoming more common in newspapers and magazines reporting on energy problems. *U. S. News and World Report* is regularly packed with statistical facts that supply firsthand data for students to interpret.

Reporting: Students in teams can investigate aspects of petroleum and gas production and consumption and report their findings back to the class. Topics can include:

☐ What happens at an oil and/or gas refinery?
☐ A deep water port and how it affects the environment.
☐ The Alaskan pipeline.
☐ Why are some people concerned about offshore drilling?
☐ Prospecting for new oil and gas reserves.
☐ How microfossils are used to indicate oil-bearing rock strata.
☐ How oil and gas were formed.
☐ Uses of petroleum and natural gas.

Encourage students to make and use diagrams, charts, and graphs in their reports to the class. Student reporters tend to make a more interesting and relaxed presentation if they can gesture toward specific points on explanatory visual materials as they talk.

Film Viewing: The American Gas Association (Dept I-14, 1515 Wilson Blvd., Arlington, Va. 22209) furnishes free 35mm filmstrips and visuals that describe production and consumption of natural gas. The filmstrip *Natural Gas—Science Behind Your Burner* shows how gas gets from well to burner and comes with a flow chart, work sheets, and gas pipeline map of the United States. It is intended for grades six through nine. *The Story of Natural Gas Energy*, intended for the same age group, tells where natural gas originates, how it was formed, and how much there is. For fourth- and fifth-grade students, the Association provides *Natural Gas Serves Our Community*, a text and cardboard cutouts that trace the route of gas from well to home.

Petroleum from Sands and Shales

The dual push of large price increases and uncertain supply of imported oil has made the extraction of petroleum from oil shales and tar sands economic realities. Huge oil shale deposits are located at the juncture of Wyoming, Utah, and Colorado; large accumulations of tar sands lie in northern Alberta. To extract oil, large amounts of rock or rocklike sand are mined and crushed and oil is literally cooked out.

Because the operations present the possibility of serious environmental damage, an investigation of the mining of oil sands and shales provides mature students with an opportunity to relate geographic, geologic, meteorologic, aesthetic, biologic, and economic factors. Different teams can investigate each factor, guided by the following questions:

Geographic: Exactly where are the deposits located? Are the deposits in mountainous, rolling, or flat terrain? Will stream patterns be affected by mining operations? Will strip mining create erosion?

Geologic: What kinds of rock or sand formations contain oil? How deep are the deposits? When were the deposits formed? under what conditions? How hard is the material? How much petroleum can be extracted from a kilogram of rock? Will groundwater sources be disturbed by mining operations?

Meteorologic: How much rain or snow does the area receive? Will wetness or dryness affect the operations? Do strong winds prevail creating a dust erosion problem? Will freezing weather interrupt operations? If so, how long? Will cooking operations pollute the air?

Aesthetic: How will mining operations affect the scenic and recreational aspects of the area? Will stream and air clarity be affected?

Biologic: What plants and animals inhabit the area? How

will mining operations affect them? How will plant removal affect the soil and its water holding capacity? Can the spoil banks of used ground rock be returned to plant and animal habitats? how?

Economic: What is the cost of extraction? What is the cost of restoring the land? What new jobs are created? Should the crude oil be piped or hauled to refineries, or should refineries be built at the site? How would use of nonimported sources of oil affect the nation's balance of payments? its international relationships?

As data are collected and presented to the class, a list of pros and cons can emerge. Once again, encourage team reporters to support their oral presentations with visuals—pictures projected with an opaque projector, maps displayed on a bulletin board, diagrams distributed in mimeographed form, or charts projected with an overhead projector.

Incidentally, if you or a student can secure some oil sands or shales, do so. They would certainly liven up the proceedings.

Nuclear Power

1939—splitting the atom; 1942—the first chain reaction; 1951—power from atomic fission; 1957—first major nuclear electric power plant, Shippingport, Pennsylvania. 1972—twenty-nine nuclear power plants in the United States; at least one nuclear power plant in each of the three West Coast states and in almost every state east of the Mississippi; nuclear power supplying less than 3 percent of the power used in the United States; 1975—fifty-two nuclear power plants in operation; the nuclear power industry's goal is to provide nearly 25 percent of the nation's electric generating capacity by the 1980s, but this goal is not likely to be achieved.

As with petroleum, students investigating this burgeoning reliance on nuclear energy will not be able to conduct firsthand experiments. However, because there are numerous sources of information that students can tap, searching printed materials to find answers to questions is a feasible activity and one that can culminate in the production of a short, illustrated book.

Group your students into three-member book-writing teams with one student primarily responsible for searching sources for information, a second primarily responsible for drawing diagrams, charts, and pictures, and a third responsible for writing the book. In setting up the teams, try to make sure that all the writers or all the artists are not on one team; spread the talent throughout the groups.

You may wish to distribute a listing of nuclear energy-related topics to guide this book-writing activity. A team selects a topic from the sheet as the focus of its book. Once students have searched for answers to questions on the topic chosen, they organize their information logically, write their manuscript, print or type the edited manuscript on plain paper, draw appropriate illustrations, and bind their pages between cardboard covers. They can add finishing touches such as a thought-provoking title, an attractive illustration on the cover, and a title page with credits and publication date. Student books on various nuclear topics can be organized as volumes in a series and contributed as references to the library.

Obviously, you will have to do some preliminary work with your class before students plunge into group investigations of topics related to nuclear power. The class as a whole could write a glossary of atomic energy terms, which would be the first volume in the series. Terms such as electron, neutron, proton, chain reaction, radioisotope, fission, reactor, and gamma radiation might be included. Students can check their own explanations of these terms against the explanations given in a booklet distributed by the Atomic Energy Commission, *Nuclear Terms, A Brief Glossary* (USAEC, P.O. Box 62, Oak Ridge, Tenn. 37831).

Nuclear energy topics that can be developed into books include:

How a Nuclear Reactor Works: What fuel is used? How is the fuel assembled? How is the chain reaction moderated? What is the arrangement of the cooling system? How is the heat transferred for conversion to mechanical energy? How is the mechanical energy transformed to electricity? What safety features are employed?

Advantages of Nuclear Reactors: How does the cost of generating a kilowatt of electricity in a reactor compare to a kilowatt generated by coal, gas, or oil? How does the cost of transporting nuclear fuel compare to the cost of transporting fossil fuels? How does the space required to store nuclear fuels compare to the space required to store fossil fuels? How clean is a nuclear reactor compared to a plant run on fossil fuel? What materials are emitted from a nuclear plant as compared to an electrical generating plant run on fossil fuel?

Pollution from Nuclear Reactors: What machines, tools, or parts of the reactor are made radioactive? How are these parts taken care of? What liquids such as cooling fluids are made radioactive? What is done with them? What radioactive gases are generated? Are they released into the air? Why are many reactors located near lakes or rivers? Why are cooling towers sometimes used? How do they work? Under what circumstances is fishing in nearby water improved? Why do massive fish kills sometimes occur?

Disposal of Radioactive Wastes: How is used radioactive fuel treated? Is it salvaged for reuse? discarded? How are radioactive waste liquids and solids stored or disposed of? Why are some radioactive wastes more dangerous than others? Why is decay time of radioisotopes important? What kinds of geologic underground formations are suitable for injection disposal of radioactive wastes?

New Kinds of Nuclear Reactors: How does the high temperature, gas-cooled reactor work? Why is so much research being done on the making of a fast breeder reactor? How would a breeder reactor be cooled? How would a fusion reactor work?

Heat from the Earth

The only geothermal power plant in the United States is located near an extinct volcano in northern California, about ninety miles from San Francisco. Although called the Geysers, the vents emit hot vapor steadily, not intermittently in fountain-like jets of hot water and stream as does famed Old Faithful in Yellowstone National Park. Hot water wells are producing electricity in Waitaki, New Zealand, and near Mexicali, Mexico. Known geothermal fields in the United States are mostly in California and Nevada. Other possible geothermal reserves are in western states, particularly Washington, Oregon, and Idaho.

The source of the heat is molten rock—magma—which is still cooling about twenty miles below the earth's crust. Above the magma, porous rock contains water that is boiled by the heat below. Solid rock capping the porous layer causes a buildup of high temperature, high pressure steam. Fissures through the hard cap rock allow some steam to escape to the surface. After drillers have tapped the steam source, pipes carry the steam to spin the blades of turbines, which in turn drive generators to produce electricity.

Problems arise because some wells yield brine; sometimes the steam and condensed water contain mineral impurities that corrode pipes and turbines. The steam may also contain ammonia, boron, and sulfur dioxide, which pollute the air or nearby streams and groundwater.

Aspects of geothermal power provide viable topics for students to investigate, such as the discovery, locations, and uses of world geothermal fields; relationships among geysers, hot springs, volcanoes, and geothermal heat; the technology of producing power from geothermal heat sources; future importance of geothermal energy in the total power market; and pollution abatement problems. Student-made overhead transparencies can show maps of geothermal fields, cross-sectional views of geysers and tapped geothermal sources, or bar graphs of increases in geothermal energy production in recent years. A working model of a geyser—simply a coffee percolator with glass top removed and hidden behind a board on which underground strata have been painted—will add action.

An illustrated folder, *The Geysers*, which details the discovery and development of geothermal energy in California, is available from the Pacific Gas and Electric Company (77 Beale St., San Francisco, Calif. 94106). A nontechnical article in the August 1973 *Environmental Science and Technology*, "Geothermal Heats Up," explains some of the techniques and problems of extraction and provides a map showing locations of geothermal fields in the western states. "Geothermal Power," an informative article in the January 1972 *Scientific American*, contains diagrams illustrating how power plants function on geothermal power, how houses can be heated, and how steam and brine can be separated. All are excellent sources for student researchers; the diagrams make great transparencies.

Solar Power

The fossil fuels we are using at an accelerated rate today were formed over a span of hundreds of millions of years. The origins of nuclear fuels and geothermal steam and hot waters go back to the beginnings of the earth. They are nonrenewable, and once reserves are used, there will be no more.

In contrast, the energy of the sun is a continuing source of new energy for our planet. When we tap solar energy, we are tapping a renewable resource. There will be more.

On an average square meter of earth surface, 600 watts of solar energy are absorbed as heat each minute of the day. At night the earth loses into space the heat it has absorbed during the day. Can this energy be harnessed before it eventually dissipates back into space?

Direct conversion of large amounts of solar energy to electric power is still a distant goal because the technology is complicated and the capital investment is high. However, there are several unsophisticated ways that students can be introduced to the concept of solar energy.

Ask them to fasten a garden hose to an outside water faucet, fill it with water, and expose the coils of the hose to sunlight. After it has been exposed for several hours, students catch the warmed water in a pail to hand test the gain in temperature. Warmed water is pushed out of the hose by the pressure of additional water when the faucet is turned on.

More mature students may wish to quantify the heat stored. They first measure the Celcius temperature of the water filling the hose, then coil the hose in a spiral with the coils against one another, and measure in meters the radius of the circle formed. They calculate the area of the coiled surface using the formula area $= \pi r^2$. After the coiled hose of water has been exposed to the sun, students collect the water used and measure the volume in milliliters and the temperature in degrees Celcius. Since one milliliter of water raised one degree Celcius absorbs one calorie of heat, junior high school students can calculate the total heat captured from the sun and stored in the heated water:

$$\left(\begin{array}{c} \text{Water temperature} \\ \text{(Celcius)} \\ \text{after} \\ \text{sun exposure} \end{array} - \begin{array}{c} \text{Water temperature} \\ \text{(Celcius)} \\ \text{before} \\ \text{sun exposure} \end{array} \right) \times \begin{array}{c} \text{Milliliters of} \\ \text{water heated} \end{array} = \begin{array}{c} \text{Calories} \\ \text{Stored} \end{array}$$

If the coiled hose occupied one square meter, then the result is expressed as calories stored per square meter of surface. If students recorded the length of time during which the water was exposed to the sun, they can also express the result as a rate; that is, calories stored per square meter of surface per hour.

Of course, the surface area is partly hose, not just water. And if the hose is warm to start, some heat may be transferred to the water from the hose so that the net gain of heat may not entirely come from the sun during the exposure time. Help students identify these limitations of the experiment.

Incidentally, students can use plastic bags of water rather than a hose. This may be more practical in urban areas where the bags can be set outside where they will not be disturbed during exposure to the sun.

A second way students can be introduced to the concept of solar energy is through construction of a solar furnace. Edmund Scientific Company markets a plastic, square foot-sized fresnel lens, with directions on how to make a solar furnace, for about $7. The fresnel lens acts as a giant magnifier that will produce a temperature of 2000° F. on a focus spot. A newspaper will burst into flame in less than a minute when solar heat is focused on it with the lens. Or use a simple magnifying lens to focus the sun's rays on a very warm day; a brown, burnt spot will appear on a piece of paper.

Another way to "see" the power of solar energy is with a radiometer. This instrument has a partial vacuum within a glass globe in which four vanes spin on a needle-point bearing. Set it in direct sunlight; it will spin at the rate of 3,000 revolutions per minute. Edmund Scientific markets a Crooke's radiometer for about $5.

Solar cells convert sunlight to electricity. They are used in weather satellites and Skylab instruments. Edmund Scientific sells solar cells ranging in price from a few dollars and miniature motors to run off them. A *Solar Cell Experimenter's Handbook* is available for less than $1. Frey Scientific markets a solar power supply with two solar cells mounted on a wood base that will operate volt meters or small motors. The outfit costs about $15 and comes with a teacher's guide.

Once students have worked with lenses, cells, radiometers, and coils of water, some may opt to construct a working model of a solar heating plant that could operate in a small home. Suggest this activity as an individual or group project to girls and boys with mechanical inclinations.

Hydroelectric Power

Water power is a renewable energy resource. Through its key role in the hydrologic cycle, the sun is the ultimate source of hydroelectric power. The heat of the sun evaporates water from ocean and land surfaces. When this water falls as precipitation on regions of higher elevation, the energy from the gravity flow of water to the sea can be harnessed. Hydroelectric power is generated when impounded water is gravity-fed against turbine blades before being released into a stream below a dam. Dams have been constructed across many of the world's rivers. There are, however, few additional river locations in the United States that from an engineering standpoint could be damned for hydroelectric power unless the remaining scenic wild rivers are converted into placid lakes.

The conflict between environmental values and energy needs is one that sophisticated upper-grade students can investigate and debate. Conservation organizations such as the Sierra Club (1050 Mills Tower, San Francisco, Calif. 94104), which has over thirty-five chapters across the country, and Friends of the Earth (30 East 42nd St., New York, N.Y. 10017) produce literature advocating preservation of the natural environments. Other concerned organizations such as the Wilderness Society (729 15th St., N.W., Washington, D. C. 20005) are described in *The Complete Ecology Fact Book* (see bibliographic entry) and are listed in the National Wildlife Federation's *Conserva-*

tion Directory. Local chapters of these organizations may be able to furnish you with printed matter and speakers. To obtain the opposing point of view, contact power generating companies, which are hard pressed to meet the energy demands of the nation.

Upper-grade students can experiment to determine relationships between shape, size, and positioning of turbine blades and the distance and mass of water fall. Investigators construct turbine blades of varying designs, perhaps starting with a most familiar and simple kind—the waterwheel, found in the past along rivers in industrial areas and used to convert the power of falling water into the mechanical and electrical power necessary to run the mills of the nation's developing industry during the 1800s and early 1900s.

Investigators can vary the angles at which water hits the turbine blades. They can try positioning the turbines horizontally as is done in one type of hydroelectric plant called a high-head plant and then vertically as is the case in low-head plants. *Van Nostrand's Scientific Encyclopedia* (Princeton: D. Van Nostrand, see most recent edition) as well as some more general encyclopedias supplies descriptions of the distinctions between these two kinds of plants. Investigators who have experimented with turbine blades held in different positions will find it interesting to read about these distinctions.

Incidentally, a classroom faucet with rubber tubing attached is a readily available source of falling water power. Students working at home on an individual basis can use water falling from a garden hose held at different heights from the turbine blades.

Student model builders will find constructing a model of a hydroelectric plant at a dam site a worthwhile project. A model will include the dam itself, impounded water, the hydroelectric superstructure, pipes to lead off water for irrigation, and even silt buildup. A cross section may show turbines and generators.

Although in most cases, a model will have no working parts, it will help students develop an understanding of water pressure and energy conversion and foster discussion of related factors such as silting and filling of impounded lakes, changes faced by living organisms as streams are converted to lakes, flood control by holding and releasing waters, irrigation possibilities, impounded lakes used for recreational purposes, and destruction of scenic, wild areas.

A visit to a hydroelectric dam is a terrific followup, if there is one nearby. The Tennessee Valley Authority system of multi-purpose dams receives visitors from all over the world. TVA, Hoover Dam on the Colorado River, and Grand Coulee Dam in Washington provide guided tours of their hydroelectric generating facilities. Allegheny Power, American Electric Power, Carolina Power and Light, Central Hudson Gas and Electric, General Public Utility, Idaho Power, Portland Gas and Electric, Pacific Power and Light, Pacific Gas and Electric, Montana Power, and Niagara Mohawk are just a few of the public utilities that employ hydroelectric power. Check your local utility to see if it has a hydroelectric plant that your students can visit.

One excellent source students can tap in investigating hydroelectric power is an article entitled "The Columbia River, Powerhouse of the Northwest," published in the December 1974 *National Geographic*. The seventeen existing and proposed dams along the Columbia River are shown on a colored relief map, and the problems of a multiple-use dam are discussed in detail.

Power from the Wind and the Tides

Writing in the November 1973 *Smithsonian* magazine, Wilson Clark proposes: "Take the ancient idea of windmills, add new designs and storage techniques, change life-styles, and you have ample electricity." According to Clark, by 1850 windmills in America contributed about 1.4 billion horsepower-hours of work (horsepower-hours $\times$ .746 = kilowatt-hours); 11.8 million tons of coal would have had to be burned to produce an equivalent amount of energy. Only twenty years later, the number of kilowatt-hours produced by windmills had been cut in half due to growing reliance on the steam engine. The windmills that had previously dotted the landscape of the plain states disappeared as electrical power lines crisscrossed the continent. In Holland, on the island of Aruba, on some of the islands of the Aegean, and in a few other locations the windmill still remains, more often a decorative feature than a power conversion one.

Then too, clipper ships powered by the wind once sailed the seas. As with windmills, wind-driven ships have all but disappeared from the seas, one remnant being the pleasure craft—the sailboat.

Of necessity, the world may now be experiencing a resurgence of interest in wind power. In the planning stage at the University of Hamburg in West Germany is a dyna-ship—a four-masted clipper ship with sails trimmed by remote control. Civil engineer professor William Heronemus of the University of Massachusetts suggests placing a 150-mile-long string of wind-driven generators along the Atlantic coastline, either on steel towers or giant offshore buoys; such wind generators would have blades about ten meters in diameter. NASA's Langley Research Center in Hampton, Virginia, is testing a vertical-axis windmill. Near Sandusky, Ohio, NASA is testing a 100-kilowatt wind machine.

Students can search out information about the use of wind power, especially its important role in the development of the plain states. Students with an engineering bent can investigate the workings of big sailing vessels, perhaps culminating their investigation with the construction of a model of a clipper ship. Budding engineers can explore the use of gears and cams to translate the circular motion of windmill blades to that of a vertical shaft and rocker arm pump. Of course, younger students can construct paper pinwheels to hold in the wind to see a concrete example of wind power at work and can read about the Netherlands to find out the purposes to which wind has been put in that country and the reasons why wind has proved particularly adaptable as an energy source there.

Young people will ask the obvious question, "What happens when there is no wind?" This problem plagued the homesteaders on the plains as well as the sailors aboard the clippers and is a challenge for student study. Solutions proposed today include charging storage batteries, tieing systems of many wind-powered generators into utility power networks so that coal or other fuel provides power during periods of slack wind, and using wind generated electricity to release hydrogen from water. Hydrogen separated from oxygen by the electrolysis of water is itself a fuel, one that can be stored for later use.

Obtaining large amounts of power from the tides, on the other hand, is probably an unrealistic expectation. There are few places, such as the Bay of Fundy on the United States-Canadian border where the tidal difference is about fifteen meters, in which trapped rising and falling tidal flow builds a head of water large enough to push turbine blades. Searching out information about tidal power, therefore, will produce fewer ideas for student investigators than will a study of wind power.

Energy Stored in Plants

A very small but tremendously significant amount of the sun's energy is trapped in plants through the process of photosynthesis. All food, fibers, and wood are synthesized through this capture. If managed properly, forests can provide an easily renewable energy resource.

There are still parts of the world where wood is burned extensively to produce heat. Instead of direct burning the destructive distillation of wood can produce combustible gases and other useful by-products.

The destructive distillation of wood can be demonstrated using the same apparatus pictured on page 191 for the destructive distillation of coal. Half fill the pyrex tube with dry wood shavings, sawdust, or broken wood splints. Heat the wood, collect the water condensate and wood tars in the bottle, and burn the released gas at the jet tip.

Test the wood tars in the bottle with Hydrion pH paper or blue litmus. The pH paper will register acid, and the litmus will turn red, indicating the presence of acid, in this case acetic acid. The gas at the jet tip contains mostly carbon dioxide and carbon monoxide with some hydrogen and methane. The water condensate contains a little methanol known as wood alcohol.

The gas mixture obtained from burnt wood has had little commercial value; however, a new furnace, which limits the oxygen so that the fuel is not completely burned, could convert plant material—especially forest and agricultural wastes—as well as municipal trash and even sewage sludge into methanol. Methanol can be used as a fuel for generating electricity and in a methanol-blend gasoline.

The new furnace is an interesting topic for student investigation for it can help both to ease the energy shortage and to aid in solid waste disposal. The *New York Times* of April 21, 1973, pictures and explains the furnace as does the December 28, 1973, issue of *Science*.

A related topic for student reporting is use of methanol as a fuel or fuel additive. An automobile mechanic, automotive engineer, or refinery fuel test technician can serve as a resource for a student seeking information.

Garbage, the Cinderella Fuel

"Garbage, the Cinderella Fuel" is the ironically appropriate name given to a report on trash-to-energy conversion by Gene

Smith in the February 24, 1974, *New York Times*. According to this account, the vast amounts of recoverable metals and the opportunity to release several hundred million barrels of low sulfur oil per year coupled with the dramatic increase in the costs of coal and oil have attracted large corporations into the waste-energy conversion field.

Conversion plants in Brockton-East Bridgewater, Massachusetts, and in St. Louis are now functioning. Saugus, Massachusetts, will shortly have one of the largest waste-to-energy plants. Other major projects are planned for Dade County and Ft. Lauderdale, Florida, Orange County, California, Hempstead, New York, Boston, Cleveland, and Bridgeport, Connecticut.

Students living in these areas may investigate aspects of the conversion process such as: What materials are or will be separated? How will the separation be carried out? How is the organic matter converted into fuel? Does the conversion plant produce odors or release particulates into the air? What districts or towns send their garbage to the trash-to-energy plant? What is the cost of producing the fuel?

Students who conduct this type of investigation may wish to report their findings to the class on self-made filmstrips. Clear filmstrips on which they can type or print short explanatory sentences and draw accompanying sketches with a wax pencil-crayon or flow pen are now available commercially. Scholastic Magazines, Inc. (906 Sylvan Ave., Englewood Cliffs, N. J. 07632) markets a Draw Your Own Filmstrip and Slide Kit, which includes transparent film, special pencils, labels, storage cans, and 35mm slide frames, for under $15. Refills of filmstrip material are available for under $5. As students explain how the conversion of trash to energy is carried out, they can flash their explanatory filmstrips on the screen. By the way, a filmstrip works particularly well as a means of reporting on any process that has several steps. Each step can be described briefly in one sentence on a frame; the next frame can be a sketch that shows the process visually.

ANALYZING ASPECTS OF ENERGY PROBLEMS

Doing the World's Work

Per capita usage of energy varies considerably from country to country. In *Science*, December 28, 1973, Luther Carter states that in none of the European countries do people consume even half as much energy as Americans. The yearly per capita consumption of electrical energy in the United States between 1950 and 1968 rose from 2,000 to 6,500 kilowatt-hours.

Interpreting charts, plotting data on graphs, and comparing graphed information can help young people comprehend the variations in energy consumption throughout the world and the grand scale on which Americans consume energy. Present students with the following two tables of statistics—one giving the population of a number of countries and a second the amount of petroleum consumed daily by each nation.

	BARRELS OF PETROLEUM CONSUMED DAILY 1972*	POPULATION ESTIMATES MID-1971**
Argentina	400,000	24,700,000
Australia	300,000	12,800,000
Britain	2,100,000	56,300,000
Canada	1,600,000	21,800,000
France	2,200,000	51,500,000
India	400,000	569,500,000
Italy	1,900,000	54,100,000
United States	15,600,000	207,100,000
Venezuela	200,000	11,100,000
West Germany	2,900,000	58,900,000

* According to Royal Dutch/Shell Group and reported in *New York Times*, Jan. 27, 1974.
** Estimates from the United Nations.

Students can convert the tabular data into a bar graph in which two bars, one for population and one for daily oil consumption, represent the data for each nation. They can also determine yearly per capita consumption of petroleum by dividing the total national daily use by the population figure and multiplying by 365. A line graph with yearly per capita use on the y or vertical axis and the nations on the x or horizontal axis can be drawn to depict differences in per capita consumption among nations.

Students working in groups can consider reasons for the differences in energy consumption by identifying the tasks for which high energy-consuming nations use electrical or mechanical energy and which low energy-consuming nations do with human or animal power or do without. They then identify living conditions that may accompany low energy consumption, e.g., less available food, fewer conveniences, and less long distance traveling.

McGraw-Hill Films (1221 Avenue of the Americas, New York, N. Y. 10020) distributes some excellent films that show how people in different parts of the world grow their food and tend to other basic needs. Films from the McGraw-Hill Intermediate Geography Series are directed to young people in upper-elementary grades; similar films for junior high students are available. Show a film that depicts the ways people in low energy-consuming nations do their work and then a film about a high energy-consuming nation such as the United States. Viewers can list differences and draw comparisons. The following table may aid you in ordering films for rental or purchase.

A view of low energy-consuming nations.

INTERMEDIATE LEVEL	JUNIOR HIGH LEVEL
India: The Struggle for Food. 18 min., color	*In India the Sun Rises in the East*, 14 min., color

INTERMEDIATE LEVEL

North Africa: Water and Man, 16 min., color

Indonesia: A Time to Grow, 19 min., color

Guyana: South America's Newest Nation, 18 min., color

JUNIOR HIGH LEVEL

Stones of Eden (A Contemporary Film), 25 min., color

The Dayak: People of Borneo, 29 min., color

A view of high energy-consuming nations.

INTERMEDIATE LEVEL

The Great Plains: From Green to Gold, 14 min., color

The Great Lakes Area: Men, Minerals and Machines, 15 min., color

The Central Farming Region: Food for the Nation, 14 min., color

JUNIOR HIGH LEVEL

France: New Horizons, 20 min., color

West Germany: the Industrial Giant, 19 min., color

Discussions of ethical questions are a next step with mature students. Is it morally right for any country to consume a high percentage of the world's limited energy reserves? Is it right for one group to have much, another to have little? Several films are useful springboards for upper elementary and junior high discussions of relationships between the "haves" and the "have-nots." Two produced by the National Film Board of Canada and distributed by McGraw-Hill in the United States are: *Man and His Resources,* 28 min., b & w, and *To Each a Rightful Share,* 28 min., b & w. The first examines the responsibility of the "have" nations to the "have-nots," and the second is concerned with people's search for a better life. A documentary produced by NBC-TV and distributed by McGraw-Hill focuses on problems of feeding the rapidly expanding populations of some of the underdeveloped areas and examines ways of harnessing the world's resources to produce

Energy Challenges

more food—*Tomorrow's World: Feeding the Billions,* 54 min., color.

A Wayne Mitchell film distributed by BFA Educational Media (2211 Michigan Ave., Santa Monica, Calif. 90404) explores the question of why one nation with a diversified economy prospers while another dependent on a single major crop remains underdeveloped—*Economic Geography: Comparing Two Nations,* 10 min., color.

Analyzing the Issues

Current newspapers are filled with tables, graphs, editorials, cartoons, letters to the editor, and syndicated columns that supply data and comments on many different aspects of the energy situation. Working with such firsthand materials will help students develop insights and interpretational skills. For example, a graph marked "Price Changes of Venezuela Oil" in the January 27, 1974, *New York Times* dramatically depicted rapid price increases over a six-month period. Interpreting that graph in terms of data about major oil producing nations that

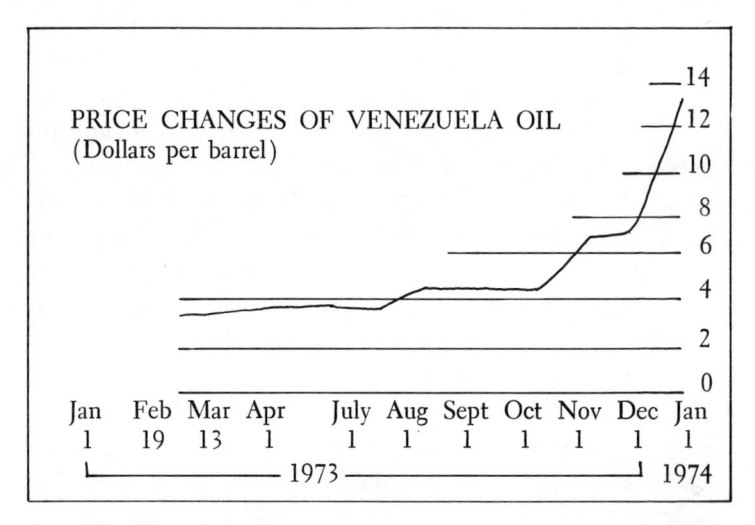

```
PRICE CHANGES OF VENEZUELA OIL
(Dollars per barrel)

Jan   Feb  Mar  Apr    July  Aug  Sept  Oct  Nov  Dec   Jan
 1     19   13   1      1    1    1     1    1    1     1
 |———————————— 1973 ————————————————————|  1974
```

Copyright 1974 by the New York Times Company. Reprinted by permission.

OIL PRODUCING COUNTRIES (Exports in millions of barrels)										
	KUWAIT	SAUDI ARABIA	IRAN	IRAQ	ABU DHABI	QATAR	LIBYA	ALGERIA	NIGERIA	VENEZUELA
1955	400	350	111	240	...	41	...	...	...	739
1962	704	589	448	348	6	68	60	...	...	1,102
1967	905	1,005	893	424	138	117	621	...	...	1,227
1969	1,006	1,158	1,159	529	219	125	1,132	331	...	1,245
1972	1,176	2,127	1,752	382	384	176	813	373	628	1,133

Copyright 1974 by the New York Times Company. Reprinted by permission.

appeared in the same issue will give students some notion of the complexity of the situation.

Students working with these tables can consider the impact of price increases on nations' balances of payment, political clout, and standards of living. They can attempt to find out which nations are large importers of the petroleum exported by the nations noted in the table. They may eventually raise the question of how can a nation like India pay for oil it needs? Or will it have to do without?

Encourage students to look through newspapers and magazines for cartoons, graphs, charts, editorials, and syndicated columns that bear on energy problems. Tack items brought to class on the bulletin board. During independent study times, young people can select a clipping from the board to analyze. Schedule a class talk-time when students can share specific information gleaned from their reading.

Political cartoons can also help develop student understanding and analytical skill. "What points about the energy situation is the political cartoonist making?" is a question you can

Energy Challenges

Copyright 1974 by the New York Times Company. Reprinted by permission.

ask your students in reference to the above or a similar cartoon.

Weekly classroom newspapers such as *Scholastic, Junior Scholastic, Search, Newstime, Young Citizen,* and *News Explorer* from Scholastic Magazines, Inc. (904 Sylvan Ave., Englewood Cliffs, N.J. 07632) and *News Parade, News Report,* and *Senior Weekly Reader* from American Education Publications (Education Center, Columbus, Ohio 43216) periodically publish charts, graphs, and cartoons related to issues in the news. As the energy crunch has become front page news, these classroom periodicals have been covering the story. Check each issue for material that students can interpret.

Conserving

Faced with evidence of rising energy costs and decreased availability of oil and gas students should begin to think about ways to cut energy consumption. Divide the conservation problem into two questions: What can we as individuals do? What can communities and industries do? Focusing on each question in turn, students can list possible actions related to lighting, heating and cooling, cooking, using appliances and machines, entertainment, transportation, and reusing and recycling materials that require large amounts of energy to produce. A big pie-shaped chart is a striking way to record actions, with things individuals can do listed at the hub and things communities and industries can do listed near the rim. The chart can easily be converted into a bulletin board. Encourage students to concoct a forceful caption for their board. Young children can add pictures clipped from magazines to a conservation bulletin board.

There are numbers of free publications available that list ways to conserve. For example, Public Service Electric and Gas (Manager, Community Relations, 80 Park Pl., Newark, N. J. 07101) distributes a brochure called "Why there is an electric and gas shortage . . . and ways you can conserve valuable energy"; your local electric company will probably be able to supply a similar piece. A & P food stores distribute a one-page flyer that lists things to do. As part of its Johnny Horizon '76 program the U.S. Department of the Interior (Washington, D. C. 20240) supplies a leaflet called "Energy, America, and You." Such items are helpful resources that students can use to check their own lists for completeness.

By the way, a related research activity is to check magazine ads and sales catalogs for unnecessary, energy-consuming devices such as the electric swizzle stick kind of gadget. These ads can be clipped and made into a chart that asks "Do We Need These Things?"

Investigating Local Energy Consumption

A local energy consumption study is one way to involve

young people directly in problems of energy use. Investigators can check the following facilities:

☐ Supermarkets, department stores, small shops, office buildings, schools, factories, and their own homes to find out: What is the inside temperature during the day? the night? What artificial lighting is used inside during the day? the night? What artificial lighting is used outside during the day? the night? What kinds of electrical machines and appliances are in use? How often are these machines used?
☐ Street lighting to find out: At what hour do street lights come on? go off? How prevalent is street lighting? In areas where thare are no street lights, do people use outside home lighting as a substitute?
☐ Recreational facilities to find out: What recreational facilities are there in the community that consume energy? What lighting, heating/cooling, or power machines are used by recreational facilities? (Note: The Houston Astrodome during 1971 used enough electricity to service more than 8,000 homes during the same period.)
☐ Transportation facilities to find out: How do citizens travel to their work? to school? to stores? to recreational facilities? What public transportation facilities exist in the community? To what extent are they used?

Besides making spot checks of public buildings, street lighting, and recreational and transportational facilities, students can question shop managers, local governmental officials, school administrators, and so forth as to what is being done locally to conserve energy. Facts and opinions gathered can be used to assess people's attempts to conserve. Students can judge whether residents are making definite efforts to turn off unneeded lights, turn down their thermostats, limit auto travel, and use public transportation where possible.

In such a survey, students may well unearth concerns. Are

people afraid that a cutback on external lighting will trigger increased crime? Do people think their jobs may be affected by a cutback in energy consumption? Are owners of restaurants, motels, and recreational facilities concerned that a gasoline shortage may affect their profits? To discover such related factors, students who are surveying opinions may wish to include the general question, How is the energy shortage directly affecting your way of life? your prosperity? your sense of safety?

This is an ideal time for students to produce a class newsletter. The results of the class's energy consumption survey, an assessment of attempts to conserve, and descriptions of people's concerns can be compiled into a dittoed newsletter. Enlivened with cartoons, riddles, jokes, and man-on-the-street interview material, the newsletter can be sent to community officials, industrial leaders, and parents.

It's Controversial!

Because issues related to energy conservation are often highly controversial, pro and con discussions are a natural in a study of energy problems. Students can informally debate the pros and cons of year-round daylight savings time, a shortened work week, changing the school calendar to eliminate days of high heating and cooling energy consumption, gas rationing, no Sunday driving, cutting back on decorative lighting such as Christmas displays, allowing use of high sulfur, air polluting fuels, limiting heating oil to 85 percent of the previous year's consumption, or changing tax laws that benefit oil companies.

Divide the class into four-person groups. Two students from each group work together to identify the pros of an issue; two identify the cons. One student from each pair states the team's position while her or his partner presents a rebuttal to the opponent's points. In this way if you have an average-sized class of thirty-two, your students can present the pros and cons of eight issues. Don't, however, schedule all debates for the same afternoon. Spread them over a few weeks to keep interest high.

Encourage students to write up their opinions and facts on an issue. Statements can be included in the class newsletter.

Meet the Press

Who is to blame for the oil shortage? The government? Big oil companies? Conservationists? The American people?

The oil companies blame government, environmentalists, and even energy consumers, setting forth their position in public relations brochures such as "Balancing Regulation and Energy" distributed by Texaco (Public Relations Division, 135 East 42nd St., New York, N. Y. 10017) and in advertisements such as Mobil's "An open letter on the gasoline shortage . . ." that appeared in 396 daily newspapers on July 7, 1973. The consumer tends to blame both government and the oil companies whose recent high profits make them suspect; "Snake Oil from the Oil Companies" in *Consumer Reports*, February 1974, analyzes the position of the oil companies and questions oil company advertisements. Governmental representatives tell consumers to temper energy demands, and they plunge into investigations of the oil companies in an attempt to assess causes for the present situation, a situation scientists have been predicting for a number of years.

But as William Smith pointed out in the *New York Times*, January 4, 1974, probably everyone is to blame—"the oil companies for being too greedy; the environmentalists for being too unbending and narrow in their viewpoint; the consumer advocates for ignoring basic laws of economics; the average American for using too much energy and wasting an unrecoverable resource;" and political leaders "for having failed to foresee that politics could supersede economics in the world's energy marketplace and for having failed to develop contingency plans . . . ."

Mature students can begin to encounter some of the complexities of energy problems in a role-playing activity. They can choose to be oil company executives, consumers, governmental officials, environmentalists, or foreign heads of state in nations having large fuel reserves and orally assess the current oil situation. To prepare for a role-playing episode, students can read publicity brochures and advertisements sponsored by oil companies, magazine articles such as "The Search for Tomorrow's Power" in the November 1972 *National Geographic* and "Snake Oil from the Oil Companies" in the February 1974 *Consumer Reports*, newspaper reports, and statements by public officials and representatives of the oil producing nations. They can record their remarks on tape, listen to their statements, and re-record until their presentations are clear and forceful. Don't forget costumes! To help get into the feel of the role they are playing, they can dress up to suit their parts.

Set this activity up as a TV "Meet the Press" session in which a role player first presents his or her views and then is questioned by members of the press—other students in the class. Schedule a number of sessions in which different parties in the controversy meet the press.

Working and Writing for Change

Young people who have studied the pros and cons of an issue and/or the positions of environmentalists, oil companies, and consumers regarding fuel shortages can formulate their own opinions. They can express these opinions in the form of letters to political leaders, consumer representatives in government, and oil executives, letters to the editor of local newspapers or national news magazines, editorials in their own classroom newsletter, or in cartoons.

In addition, they can be encouraged to keep an ongoing diary documenting ways in which energy shortages are affecting them and actions they are taking to conserve. The natural next step is for young people to write about how they personally react to energy challenges.

Student diaries compiled over several weeks can be the source material for a "You Were There in the Energy Crunch" mock radio broadcast. Students draw from their diaries to compose a script for the "You Were There" program, rehearse their lines, and record their program on tape.

Writing an energy fact book is another activity in which student investigators can summarize facts they have uncovered. As a class, students brainstorm words beginning with A that relate to energy production or consumption and then do the same for the successive letters of the alphabet. A class reporter records on paper or the board words that classmates call out. A list of energy-related words is given below so that if students find it difficult to think of a word for a particular letter, you can help out by contributing a word or two. Don't take away the fun of brainstorming, however, by giving students a ready-made list. Even primary-grade children can supply their own words.

After developing a list of words, students form small writing-illustrating groups. Each group selects a word for each letter of the alphabet, writes a statement about it, and then illustrates the word by locating pictures in magazines or making original sketches. The statements and pictures are then collated into an *ABC Energy Fact Book*. One group we know compiled its statements and sketches into something they called *An Energy Encyclopedia*.

LET'S WRITE AN ENERGY FACT BOOK

A   atomic energy, Arabs, anxiety, automobiles, airplanes, appliances, Appalachia

B   barrels of oil, buses, BP, blackouts

C   coal, coke, cars, consumers, crisis, crunch

D   deposits of oil and gas, diesel engine, drilling

E   energy, engines, electricity, ecology, environment

F	fuel, frantic buying, furnaces
G	geothermal, gas, gallons, gas stations
H	hydroelectric power, heat
I	internal combustion engine, incinerators
J	jets, jobs, judgment day, jury, Jurassic, justify, justice
K	kerosene, kite (the way it got started), kitchen, kilowatt, kilowatt-hour, Kuwait
L	light, liquid gas, lightning, long lines
M	motors, machines, mechanical energy
N	natural gas, nature, North Sea
O	oil, oil sands, oily shales, oil spills, offshore drilling
P	petroleum, producers, people
Q	quit washing, quest, quay, quality of air
R	radiometer, railroads, rigs
S	solar energy, steam-driven engines, strip mining, sheikdom, supertankers
T	tides, trains, turbines, Texas, tankers
U	uranium, underwater deposits
V	vacuum cleaners, vote, vow, volcano, volts, voltage
W	wind, water, waves, waste, windfall profits, watt
X	X-out energy waste, X-rays, X-roads. (E)xxon and the other oil giants
Y	you, young people, youth, your responsibility
Z	zero energy, zest for life, zero hour

The Energy Song

Pathmark Stores in the Northeast are responsible for a great little song that is sung by the Newark Boys Choir as part of a Pathmark TV commercial. Students involved in energy investigations may enjoy learning and singing the song. It is reprinted below with permission of Pathmark.

KEEP EARTH CLEAN, BLUE AND GREEN

CHAPTER 8

A PARTING THOUGHT

On September 3, 1802, William Wordsworth stood upon Westminster Bridge in the predawn, looked down on the London spread before him, and wrote:

> Earth has not anything to show more fair:
> Dull would he be of soul who could pass by
> A sight so touching in its majesty:
> This city now doth, like a garment, wear
> The beauty of the morning; silent, bare,
> Ships, towers, domes, theaters, and temples lie
> Open unto the fields, and to the sky;
> All bright and glittering in the smokeless air.
> Never did sun more beautifully steep
> In his first splendor, valley, rock, or hill;
> Ne'er saw I, never felt, a calm so deep!
> The river glideth at his own sweet will:
> Dear God! the very houses seem asleep;
> And all that mighty heart is lying still!

Contrast Wordsworth's London with the London of the late 1900s. Still a city of considerable majesty—of ships, towers, domes, theaters, temples—the city bears the scars of industrial progress: smoky air, a foul river, few open fields, little calm, a dense population, and a way of life totally dependent on electricity.

Writing in the November 1972 *National Geographic*, Kenneth Weaver described the situation faced by Londoners when a miners' strike reduced coal stocks, the source of most of Britain's electricity.

Minutes ago the lights flickered, went out briefly, snapped on again. It was a warning. The electricity would last only a few moments longer, and then we would be plunged into three hours of darkness.

Now I am writing by the light of candles . . . . Outside, no street lights glow, no storefronts blaze, no traffic signals wink. Only the occasional flash of automobile headlights relieves the six o'clock gloom of this February evening. . . .

In the past three days I have covered miles of London streets, seeing what happens when modern man loses the electricity which he takes for granted, and on which his material civilization is based. . . .

Along busy Oxford Street, shops and restaurants had turned off all their display lights for the duration. Inside, candles and pressure lanterns filled in during the blackouts . . . .

Piccadilly Circus, usually a dazzling glitter of advertising signs lay in murk except for lights over subway entrances. . . .

As the three-hour blackout periods shifted from area to area and back again, people across the land fretted about whether frozen foods would spoil; dairies adjusted to hours when power would be available for milking machines; factories went on part-time schedules . . . . Houses chilled, refrigerators warmed, elevators halted, and factories slowed.

Weaver found the London blackout "sobering" and considered it "a preview of things that could happen elsewhere," especially in the United States "where six percent of the world's population uses thirty-five percent of the world's energy." Weaver's prediction may be coming true sooner than even he anticipated.

To teach young people to assume some responsibility for the wise consumption of energy and for the care of our air, water, and soil environments is a matter of the utmost importance. It is a task that cannot be put off till tomorrow, for tomorrow has become today.

ANNOTATED RESOURCE BIBLIOGRAPHY

TEACHER REFERENCES

BRODINE, VIRGINIA. *Air Pollution.* New York: Harcourt Brace Jovanovich, 1973.

A general book beginning with a specific case—Episode 104—that makes both for interesting and informative reading; illustrated with graphs, charts, and black and white photographs.

Conservation Directory. Washington, D. C.: National Wildlife Federation, annual.

A list of organizations, agencies, and officials concerned with natural resource use and management in the United States.

DETWYLER, THOMAS R. *Man's Impact on Environment.* New York: McGraw-Hill, 1971.

Selected, sometimes technical articles on all aspects of the relationships between people and their environment: causes, impact on atmosphere and climate, water, land, and soils; the destruction of vegetation and animal life, and the spread of organisms.

DURRENBERGER, ROBERT W. *Dictionary of the Environmental Sciences.* Palo Alto, Calif.: National Press Books, 1973.

An illustrated dictionary for novices and professionals that covers a broad range of areas including anthropology, botany, economics, geography, geology, pollution, soils, and zoology.

EMMEL, THOMAS C. *An Introduction to Ecology and Population Biology.* New York: Norton, 1973.

A concise, clear development of basic principles of ecology and a survey of major plant communities, environmental alterations and pollution, population growth, and planning.

Environmental Education Evaluation Project. *Annotated Bibliography and Directory.* P. O. Box 2631, Toledo, Ohio 43606.

A listing of more than 200 free and inexpensive publications for environmental education with grade levels indicated. Write Dr. Jerry Underfer for information and price.

FAGAN, JOHN J. *The Earth Environment.* Englewood Cliffs, N. J.: Prentice-Hall, 1974.

A clearly written, nontechnical book explaining facts and principles behind current environmental problems and focusing on air, water, energy, oceans, soils and food, and resource limits; useful black and white photographs and line diagrams; good for upper-grade students as well.

FEGELY, THOMAS, RITA REEMER, AND LYNN RINEHARD. *Recycling.* Emmaus, Pa.: Rodale Press, 1973.

An attractive paperback in magazine format explaining recycling and the need for it, providing activities for class or individual use, and listing glass and metal recycling firms across the continental United States.

GARVEY, GERALD. *Energy, Ecology, Economy.* New York: Norton, 1972.

Chapters on depletive waste, energy and ecology, environmental costs of coal, petroleum spills, urban air pollution, nuclear power, water quality, the role of technology; central theme is that the solution of energy-associated environmental problems will require timely responses based on facts and a mix of technological and policy adjustments.

How Can Our Physical Environment Best Be Controlled and Developed? 91st Congress, 2nd session, document 91-66. Washington, D.C.: Government Printing Office, 1972.

A compilation of pertinent excerpts and bibliographic references in which opposing points of view are expressed.

McCaull, Julian, and Janice Crossland. *Water Pollution*. New York: Harcourt Brace Jovanovich, 1974.

A clear discussion of specific aspects of water pollution: chemical pollution, thermal pollution, eutrophication, preventive measures, and so forth; illustrated with sharp black and white photographs.

Mason, Joan. *Paper Making as an Artistic Craft*. London: Faber and Faber, 1959.

A clear, detailed account of the author's practical experiences making paper, including historical background of paper making in other lands; illustrated with sketches.

Miles, Betty. *Save the Earth! An Ecology Handbook for Kids*. New York: Knopf, 1974.

Suggestions on how to study and monitor pollution; actions children, schools, and parents can take to pressure industry and government to start protecting our environment.

Murdoch, William W. *Environment: Resources, Pollution, and Society*. Stamford, Conn.: Sinauer Associates, 1971.

Twenty-one authoritative analyses of the problems, with chapters on mineral, energy, food, land, and water resources; on air, fresh water, ocean pollution, ionizing radiation, pesticides, and weather-climate; and on the urban environment, economics, law, and administration; highly readable.

Nobile, Philip, and John Deedy. *The Complete Ecology Fact Book*. Garden City, N.Y.: Doubleday/Anchor, 1972.

A compilation of facts and figures on population, endangered animal species, air, water, nuclear and noise pollution, detergents, food supply, pesticides, minerals, and solid wastes; brief summaries of notable court decisions on environmental problems, lists of organizations, governmental agencies, and congressional committees having environmental concerns.

Sunset Guide to Organic Gardening. Menlo Park, Calif.: Lane Books, 1971.

Illustrated hints on practical gardening using an organic approach; topics such as how to improve the soil structure, phosphorus, nutrients cycle, preparing soil, how to start seeds, and so forth; one of a series of inexpensive how-to-do-it books found in book and magazine outlets.

TURK, AMOS, JONATHAN TURK, AND JANET TURK WITTES. *Ecology, Pollution, and Environment*. Philadelphia: Saunders, 1972.

Discussion of aspects of environmental concern including agricultural disruptions, radiation, solid wastes, thermal pollution, and noise; relevant physical science background material presented where needed; problems presented not as science but as social issues; illustrated with useful photos, diagrams, and political cartoons.

U. S. DEPARTMENT OF THE INTERIOR BUREAU OF MINES. *Minerals Yearbook*. Washington, D.C.: Superintendent of Documents, GPO, annual.

A complete source of statistical data on natural gas, coal, petroleum, petroleum products, and natural gas liquids that students can graph.

U. S. DEPARTMENT OF THE INTERIOR. *Mineral Facts and Problems*, Bureau of Mines Bulletin #650. Washington, D.C.: Superintendent of Documents, GPO, 1970.

Data on coal, energy, petroleum.

U. S. DEPARTMENT OF THE INTERIOR. *River of Life, Water: The Environmental Challenge*. Washington, D.C.: Superintendent of Documents, GPO, 1970.

One of the conservation yearbook series, presenting the story of the hydrologic cycle and how we have mismanaged water, showing water-rich and water-poor areas of the United States, and discussing the creation of more fresh water, fish, and wildlife; lavishly illustrated with colored photographs.

WAGNER, RICHARD H. *Environment and Man*, 2nd ed. New York: Norton, 1974.

An uncomplicated environmental textbook, emphasizing basic environmental problems such as man in the environment, soils, fire, water pollution, air pollution, radiation and nuclear power, pesticides, inorganic pollutants, man's urban environment, and population growth and control; indexed.

BOOKS FOR YOUNG PEOPLE

BLOOME, ENID. *The Water We Drink.* New York: Doubleday, 1971.

A discussion of uses of water—drinking, bathing, swimming, boating—and evidences of pollution; illustrated with black and white pictures; written in short sentences; lower elementary.

BRANLEY, FRANKLYN. *High Sounds, Low Sounds.* New York: Crowell, 1967.

An explanation of relationships between vibrations and sounds; illustrated with cartoonlike pictures; lower elementary.

BUSCH, PHYLLIS S. *Puddles and Ponds.* Cleveland: World Publishing, 1969.

A picture book of black and white photographs showing some of the plants and animals found in and around a pond; lower elementary.

ENVIRONMENTAL ACTION COALITION. *Eco-News.* 235 East 49th St., New York, N.Y. 10017.

A monthly newspaper published for children. Write for full description and subscription price.

EVANS, EVA K. *The Dirt Book.* Boston: Little, Brown, 1969.

Elementary explanations of soil, animals living in the soil, how soil supports plants, and uses of sand and clay; illustrated with simple sketches; lower elementary.

FREEMAN, IRA. *Science of Sound and Ultrasonics.* New York: Random House, 1968.

A description of sound waves and how sounds are made by various musical instruments; good photographs; upper elementary.

GABEL, MARGARET. *Sparrows Don't Drop Candy Wrappers.* New York: Dodd, Mead, 1971.

A short book that carries the message that people are responsible for pollution; illustrated in beautiful pen and ink drawings and written in a simple vocabulary that make the reader feel as if a child had written the book; lower elementary.

HAHN, JAMES AND LYNN. *Recycling: Reusing Our World's Solid Wastes.* New York: Franklin Watts, 1973.

A straightforward discussion of ways paper, glass, metal, and garbage are recycled; black and white photos as well as a comprehensive glossary of recycling terms; upper elementary.

HALACY, D. S., JR. *The Energy Trap.* New York: Four Winds Press, 1975.

An explanation of the energy shortage that the developed nations and particularly the United States are facing. Directions we must go are suggested and we are urged to avoid patchwork measures and "business as usual" attitudes; middle school and up.

———. *Now or Never: The Fight Against Pollution.* New York: Four Winds Press, 1971.

A straightforward discussion of the many forms and problems of pollution that result from careless affluence. Filled with specific journalistic anecdotes that students can excerpt; middle school and up.

———. *The Water Crisis.* New York: Dutton, 1966.

A discussion of the scientific aspects of the water crisis, of people's dependency on water, and of ways of managing the water supply; mature upper elementary and junior high.

HILTON, SUZANNE. *How Do They Get Rid of It?* Philadelphia: Westminster Press, 1970.

An interesting description of man's ingenuity in reusing and recycling items; upper elementary.

HYDE, MARGARET O. *For Pollution Fighters Only.* New York: McGraw-Hill, 1971.

Short chapters on different kinds of pollution and what a young person can do; hints for pollution fighters; extensive listing of federal, state, and private organizations concerned with pollution control or abatement; mature upper elementary and junior high.

KANE, HENRY B. *The Tale of a Pond*. New York: Knopf, 1960.
A year's cycle in the life of a pond told by a boy seeing the myriad organisms and their interactions; well illustrated with photos and sketches; details explained in an interesting way; good index; upper elementary and junior high.

KETTELKAMP, LARRY. *The Magic of Sound*. New York: Morrow, 1956.
An explanation of different ways sounds are made, including directions on how to make stage prop sounds; illustrated with black and white sketches; elementary.

KNIGHT, DAVID. *The First Book of Sound*. New York: Franklin Watts, 1960.
Brief explanations of many sound phenomena; includes a checklist of sound facts; upper elementary.

LEAF, MUNRO. *Who Cares? I Do*. Philadelphia: Lippincott, 1971.
The litter story illustrated with actual photographs and cartoons.

LENT, HENRY B. *Agriculture U.S.A.* New York: Dutton, 1968.
The story of farming across the United States—crops and animals, where they are raised, how they are processed—and of problems in rural areas; short chapter on careers in agriculture; illustrated with black and white photographs; upper elementary and junior high.

LEOPOLD, LUNA B., AND KENNETH S. DAVIS. *Water*. New York: Time-Life Books, 1966.
All about water—what it is, the earth-water cycle, underground reservoirs, erosion, water physiology in man and other living organisms, waterways of the world, water power, and making fresh water to drink; well illustrated with large colored photographs and attractive diagrams; elementary through adult.

LEWIS, ALFRED. *Clean the Air*. New York: McGraw-Hill, 1965.
A description of the fumes, particles, smoke, smog, and smaze in the air; illustrated; upper elementary and junior high.

———. *This Thirsty World*. New York: McGraw-Hill, 1964.

A description of the ways engineers are replenishing underground water sources, reclaiming waste water, searching for ways to desalt oceans, and conserving available sources; upper elementary and junior high.

McFall, Christie. *Wonders of Sand*. New York: Dodd, Mead, 1966.

Straightforward explanations of how sand is moved by wind, ocean tides, and waves, how dunes are stabilized, and kinds of sands and life found along the ocean shore; illustrated with diagrams and black and white photographs; upper elementary and junior high students, especially those living near the seashore.

Marshall, James. *The Air We Live In: Air Pollution, What We Must Do About It*. New York: Coward, McCann, 1969.

Brief, clear explanations of air pollutants and their damage and ways of controlling pollutants; illustrated with large sharp photographs and diagrams; upper elementary and junior high.

———. *Going to Waste: Where Will All the Garbage Go?* New York: Coward, McCann, 1972.

All about garbage—what it is, why there is a growing mountain of it, packaging, waste collection, what we do with waste, and what we can do to reduce the load; well illustrated with black and white photographs that can be appreciated even by the very young child; upper elementary and junior high.

Marx, Wesley. *The Protected Ocean: How to Keep the Seas Alive*. New York: Coward, McCann, 1972.

Another book in the "New Conservation" series published by Coward that answers questions such as why we need the ocean, how the ocean works, how the ocean is endangered, and how we can protect the ocean; good for students in middle and junior high grades. Other titles in the series: *The Land We Live on: Restoring Our Most Valuable Resource* by John Vosburgh, *Our Threatened Wildlife: An Ecological Study* by Bill Perry, *Man, Earth and Change: The Principles and History of Conservation* by Jean Worth, and *The Water We Live By: How to Manage It Wisely* by L. A. Heindl.

May, Julian. *Blue River.* New York: Holiday House, 1971.
The story of a real river that almost died from pollution and what people are presently doing to save it; colorful illustrations and simple sentences; lower elementary.

Pringle, Laurence. *Ecology: Science of Survival.* New York: Macmillan, 1971.
An exploration of the intricate web connecting humans to their environment and to other forms of life; upper elementary.

Rondiere, Pierre. *Purity or Pollution: The Struggle for Water.* New York: Franklin Watts, 1971.
A thorough discussion of water—its origins, properties, importance to man—and of the ways people are overloading their water supply; beautiful, full-color illustrations; upper elementary to adult.

Seuss, Dr. (Geisel, T. S.). *The Lorax.* New York: Random House, 1971.
A picture storybook that tells about the Lorax's warning to the Once-ler not to chop the trees or pollute the air and the Once-ler's failure to listen to the warning; illustrated with typical Seuss drawings; elementary.

Shuttlesworth, Dorothy E. *Clean Air-Sparkling Water: The Fight Against Pollution.* Garden City, N.Y.: Doubleday, 1968.
A short discussion of the air and water situation; illustrated with rather good black and white photographs that can be projected with an opaque projector for class viewing; grades four through seven.

Simon, Seymour. *Science Projects in Pollution.* New York: Holiday House, 1972.
Experiments students can conduct to investigate air pollution, water pollution, and pollution of the earth; upper elementary.

Sootin, Harry. *Easy Experiments with Water Pollution.* New York: Four Winds Press, 1974.
Illustrated step-by-step experiments requiring a minimum of materials. Each group is preceded by a short background chapter includ-

ing experiment hints. Topics include water purification, water softening, toxic metals in water, detergent and soap action; upper elementary.

———. *Science Experiments with Sound*. New York: Norton, 1964.

A variety of experiments using easily obtainable materials; illustrated by clear diagrams; upper elementary.

STEPHENS, J. H. *Water and Waste*. New York: St. Martin's Press, 1969.

An English book that tells about problems of water supply and sewage treatment in the London area; good for upper-grade students undertaking a case-study investigation in this area.

TANNENBAUM, BEULAH, AND MYRA STILLMAN. *Understanding Sound*. New York: McGraw-Hill, 1972.

Clear explanations of sound phenomena with interesting, practical anecdotes, especially in the chapter on reflecting, absorbing, and directing sound; upper elementary.

VARGO, DONALD J. *Wind Energy Developments in the 20th Century*. Cleveland: Lewis Research Center, National Aeronautics and Space Administration, 1975.

A description of the ways wind energy has been used by the Dutch, Danes, Russians, Americans, French, and Germans, and the work currently being done, especially by NASA; illustrated with line drawings; good readers in upper elementary grades.

FILMS FOR CLASSROOM USE

The Aging of Lakes, color, 14 min., Encyclopaedia Britannica Educational Corp., 425 N. Michigan Ave., Chicago, Ill. 60611.

An investigation of eutrophication of lakes, causes, and possible remedies. Student viewers can list causes and remedies.

Air Pollution, color, 11½ min., Journal Films, Inc., 909 W. Diversey Parkway, Chicago, Ill. 60614.

A number of facts about air pollution. Students viewers can then write their own air pollution fact book.

Alaska: Setting a New Frontier, color, 22 min., Films Incorporated, 1144 Wilmette Ave., Wilmette, Ill. 60091.

A National Geographic production documenting changes occurring in Alaska and the attempts being made to prevent the wholesale destruction of Alaska's natural beauty. Have students suggest other problems that must be solved to prevent the destruction of the nation's last frontier.

All the Difference, color, 20 min., Modern Talking Picture Service, Inc., 2323 New Hyde Park Rd., New Hyde Park, N.Y. 11040, free loan.

A poetic plea to halt the despoiling of nature in America. Use this one to trigger the writing of haikus.

Buttercup, color, 11 min., Churchill Films, 662 N. Robertson Blvd., Los Angeles, Calif. 90069.

The journey of a buttercup downstream as it encounters first the calm of the countryside and then the hostile water environment of an industrial area. Since this is a wordless film, students can write a script to accompany it and then tape the script.

Conservation: For the First Time, color, 9 min., McGraw-Hill Films, 1221 Ave. of the Americas, New York, N.Y. 10020.

A film for lower elementary students that shows how a group of children filmed their city—buildings, automobiles, birds, and trees. Motivate your students to develop a similar photographic project.

The Day Is Two Feet Long, color, 9 min., Weston Woods, Weston, Conn. 06880 or Modern Film Rental Library, 1212 Ave. of the Americas, New York, N.Y. 10036.

Brief but beautiful views of nature with a quiet musical accompaniment; advertised as a haiku experience, this film is good for stimulating written descriptions of the natural environment.

Ecology Primer, color, 18 min., American Educational Films, 331 N. Maple Dr., Beverly Hills, Calif. 90210.

A film narrated by Dennis Weaver that examines problems of pollution, waste disposal, and population growth and suggests action that young people can take. Students can put into effect some of the ideas suggested.

Estuarine Heritage, color, 28 min., Motion Picture Service of U. S. Dept. of Commerce, National Oceanic and Atmospheric Administration, 12231 Wilkins Ave., Rockville, Md. 20852, free loan.

The effects of pollution on animals that live in estuaries—oysters, clams, and shrimp. This may precede or follow the work with a clam described in Chapter 4.

The Fence, color, 7 min., BFA Educational Media, 2211 Michigan Ave., P. O. Box 1795, Santa Monica, Calif. 90406.

A wordless allegory in which a man throws trash into another's yard, setting in motion a series of acts that lead to disaster. Students can compose their own allegories by brainstorming possible story plots and then pantomime the stories in teams.

For Your Pleasure, color, 4 min., Mass Media Associates, 2116 N. Charles St., Baltimore, Md. 21218.

An environmental cartoon in which troll-like people cut the trees, pollute rivers, build cities and highways, and overpopulate. Student viewers can draw and write comic strips with the trolls as characters and an environmental theme. Followup with a dramatization of Dr. Seuss's *The Lorax*.

Garbage, color, 10½ min., BFA Educational Media or from Holt, Rinehart & Winston, Media Dept., 383 Madison Ave., New York, N.Y. 10017.

A photographic description of the masses of garbage being produced by society. Garbage is seen as art, an index to character, and a menace. Student viewers in upper grades can suggest other ways to conceive of garbage and develop metaphors and similes to describe it.

Litter Bug, color, 8 min., Walt Disney Educational Materials Co., 800 Sonora Ave., Glendale, Calif. 91201.

A Donald Duck film that illustrates several kinds of litter bugs—the beach bug, the sneak bug, the car-riding bug—and illustrates the seriousness of the litter problem by reducing it to the laughable. Use this to introduce the whole notion of litter, and have students orally describe other varieties of litter bugs they have observed. A possible follow-up is to sketch cartoons showing the varieties of litter bugs.

The Litter Monster, color, 16½ min., Alfred Higgins Productions, 9100 Sunset Blvd., Los Angeles, Calif. 90069.

A description of cleanup activities that young people in rural and suburban areas can attempt; has a rock-type theme song that young people may enjoy singing.

Oil Spoil, color, 17 min., Association-Sterling Films, 866 Third Ave., New York, N.Y. 10022, free loan.

A Sierra Club film without narration that documents the environmental destructiveness of oil and automobiles; good shots of the California coast and oil-soaked beaches. Have upper-grade students write and tape their own narration.

Pave It and Paint It Green, color, 27 min., University of California, Extension Media Center, Berkeley, Calif. 94720.

A film without narration that shows the natural beauty of Yosemite National Park and the way it has been filled with trash, people, cars, and campers. Students can write a litany as a narration for the film.

The Problem with Water Is People, color, 30 min., McGraw-Hill Films, 1221 Ave. of the Americas, New York, N.Y. 10020.

A photographic documentary of the Colorado River's journey from its mountain source to the ocean, showing instances of pollution and poor water management. An upper-grade student can write a case-study report based on the film, which is narrated by Chet Huntley.

Problems of Conservation: Air, color, 15 min., Encyclopaedia Britannica Educational Corp., 425 N. Michigan Ave., Chicago, Ill. 60611.

Scenes of smokestacks emitting red smoke and of smoggy English industrial towns. A description of some well-known cases of intensive air pollution—1952 in London, 1948 in Donora, Pa. Try a follow-up art activity in which students fingerpaint scenes of industrial air pollution.

Problems of Conservation: Soil, color, 14 min., Encyclopaedia Britannica Educational Corp.; address above.

Scenes of soil destruction through paving, removing plant cover, and overgrazing! Scenes of soil conservation through contour plowing, crop rotation, and terracing. Use this one to motivate some of the activities in Chapter 5.

Runaround, color, 17½ min., American Lung Assn., 1740 Broadway, New York, N.Y. 10019, free loan.

An animated film that follows a hero confronting a series of air polluters—electric utility executives, industrialists, auto manufacturers, incinerator operators—each of whom passes the buck. Use this one to trigger role-playing, with individual viewers later taking the part of a spokesperson for each group of polluters.

Something in the Air, color, 28 min., Modern Talking Picture Services, 2323 New Hyde Park Rd., New Hyde Park, N.Y. 11040, free loan.

A film suggesting conversion to nonfossil fueled engines: diesel, rotary, and electric engines. Students in upper grades can individually investigate some of these alternatives.

They Care for a City: Mission Possible, color, 52 min., McGraw-Hill Films, 1221 Ave. of the Americas, New York, N.Y. 10020.

A look at the way San Francisco is fighting air and water pollution, transportation, and other urban problems. Upper-grade viewers may wish to conduct a similar study of their community.

They Care for the Land: Mission Possible, color, 53 min., McGraw-Hill Films; address above.

A look at conservation and environmental problems in Florida. Upper-grade viewers may wish to identify similar problems in their own state.

Time to Begin, color, 29 min., McGraw-Hill Films; address above.

Scenes of littered roads, auto graveyards, polluted air and water in contrast with scenes of countryside and metropolitan elegance. Use this film to encourage students to draw contrasting scenes; watercolor might be a good medium for the art activity.

Tup Tup, color, 8¾ min., BFA Educational Media, 2211 Michigan Ave., P. O. Box 1795, Santa Monica, Calif. 90406.

Another film without narration—this one telling about the disturbance to a man's peace created by a persistent thump, thump. Try to follow up with a taping experience in which students make tapes of noises in the environment.

Waters of Coweeta, color, 20 min., regional offices of the Forest Service, free loan.

A vivid pictorial description of various ways of managing the forest and land: steep land left as forest, land cleared of forest, and land farmed with outmoded practices. Students experimenting with erosion and different kinds of ground cover will find this one helpful.

Work in Progress: Steel and the Environment, color, 28½ min., American Iron and Steel Institute, 1000 16th St., N.W., Washington, D.C. 20036, free loan.

A description of the programs to clean up and preserve the environment being carried on by the steel industry. Since the film fails to mention the harmful things the industry does, the film is great to use as a base for analyzing publicity statements of industries.

Yours for a Change, color, 15 min., Association-Sterling Films, 866 Third Ave., New York, N.Y. 10022, free loan.

An explanation of basic ecological concepts and a description of some projects for upper-elementary students such as making vest-pocket parks and outdoor murals. Helps young people see some of the problems they will encounter as they attempt similar projects.

COMPANIES MARKETING AUDIO AND/OR VISUAL MATERIALS

Robert J. Brady Co., Bowie, Md. 20715 (Division of Prentice-Hall Learning Systems, Inc., Englewood Cliffs, N.J. 07632).

Environment and You, an 8-book series, each with teaching objectives, lesson contents, activities, and a combination of 17-18 spirit masters, color transparencies, and posters. *Jumper's Environment, Pollution in Our Environment, Interactions in the Environment,* and *Pollution Has Many Forms* for lower grades; *Pollution and You, Pollution Is Growing, Your Future in Ecology,* and *Our Influence on the Environment* for upper elementary.

Educational Images, P. O. Box 367, Lyons Falls, N.Y. 13368.

Slide sets (20 per set) with accompanying text: *Ecology of a Pond* —inhabitants and interrelationships in a freshwater pond; *Erosion*— natural and man-caused erosion shown both on a small and a large scale. Upper elementary and junior high school.

Educational Manpower, Inc., P. O. Box 4272-A, Madison, Wis. 53711.

Air Pollution Multi-Media Set, 2 filmstrips with record or cassette, 9 transparencies, 23 overlays, a teacher's guide: the causes and effects of air pollution; related title—*Water Pollution Multi-Media Set*. Middle and junior high school.

Environmental Studies, 6 color-sound filmstrips with records or cassettes; topics: *Rivers Must Not Die; Solid Wastes; Pests, Pesticides, and People; Noise; Land and the Soil;* and *Air Pollution*. Elementary and junior high school.

Hayes School Publishing Co., Inc., 321 Pennwood Ave., Wilkinsburg, Pa. 15221.

Ecology Posters, color, 12 x 18-inch posters emphasizing concern for the environment. Elementary.

Hudson Photographic Industries, Inc., Irvington-on-Hudson, N.Y. 10533.

Naturally Speaking, 4 sound-color filmstrips with cassettes, program guide; an awareness-building package that carries the viewer to a pond and the woods. Elementary and junior high.

Instructor Publications, Inc., P. O. Box 6108, Duluth, Minn. 55806.

Eco-Problems Posters, a set of 12 16 x 25-inch posters on such topics as solid waste disposal, oil pollution, and heat pollution.

Learning Arts, P. O. Box 917, Wichita, Kan. 67201.

Ecological Crisis, 6 sound filmstrips with cassettes or records; topics: *Populaton Statistics, Population Trends, Ecological Considerations, Evolution and Extinction, Pesticides,* and *Pollution*. Upper elementary and junior high.

Ecology: The People Are Scratching, record or cassette: ecology-related songs. All levels.

Environmental Awareness, 5 sound filmstrips with records or cassettes; splendid color photographs; topics: *Colors in Nature, Textures in Nature, Patterns in Nature, Awareness in the City,* and *Awareness in Forest and Field* with a musical accompaniment. Elementary. (Available also from Educational Manpower.)

Miller-Brody Productions, 342 Madison Ave., New York, N.Y. 10017.

Color filmstrips with records or cassettes; titles: *Ecology, Can Man and Nature Coexist?*—shows massive outpouring of waste; *Conservation Through Recycling,* part 1—recycling iron and steel scrap and abandoned cars, part 2—recycling paper, rags, rubber, and metals; *Our Polluted World: The Price of Progress*—the global problem; *Energy: Impact on Values and Lifestyles*—implications of the end of an era of abundant cheap energy.

National Geographic Society, P. O. Box 1640, Washington, D.C. 20036.

Small Worlds of Life, an exceptional set of 7 color filmstrips with a teacher's guide; topics: *The Pond, The Coral Reef, The Apple Tree, The Tundra, The Salt Marsh,* and *The Everglades.* Elementary and up.

This World of Energy, a comprehensive set of filmstrips with records or cassettes; topics: *Energy in the Earth, Using Energy, Fossil Fuels, Nuclear Power,* and *Energy for the Future.* Also available boxed with teacher's guides and library cards. Intermediate and up.

National Wildlife Federation, 1412 Sixteenth St., N.W., Washington, D.C. 20036.

Environmental Discovery Units, booklets for students to use as they investigate aspects of their environment; topics: *Vacant Lot Studies, Soil, Man's Habitat, The Urban Ecosystem, Stream Profiles, Change in a Small Ecosystem, Outdoor Fun for Students,* and so forth. Elementary.

Roy G. Scarfo, Inc., P. O. Box 217, Thorndale, Pa. 19372.

The Air Pollution Chart and *The Water Pollution Chart,* full-color wall charts that make good bulletin board displays. Elementary and junior high.

Schloat Productions, 150 White Plains Rd., Tarrytown, N.Y. 10591.

Air Pollution, 2 color-sound filmstrips with cassettes or records, program guide: causes and effects of air pollution. Junior high.

The Beer Can by the Highway, 1 color-sound filmstrip with record or cassette, program guide: the problems of our throw-away culture. Junior high.

Soil: Its Meaning for Man, 2 color-sound filmstrips with cassettes or records, program guide: misuses of soil and the consequences. Junior high.

Voice of the Water, 2 color-sound filmstrips with records or cassettes, program guide: water-air cycles, food-soil production, and interdependence. Elementary and junior high.

Scholastic Audiovisual Materials, 904 Sylvan Ave., Englewood Cliffs, N.J. 07632.

Draw Your Own Filmstrip and Slide Kit, materials for handmaking filmstrips: clear film, special pencils, labels; students can use this kit to make their own visuals to accompany reports. (Similar kit available from Hudson Photographic.)

Earth Corps Program, multi-media ecology/environmental awareness units; titles for grades five and six: *Sharing the Earth*, *First Follow Nature*; related color filmstrip is *Earth: Oasis in Space*.

12-inch LP recordings: *Documentary Sounds*—noises of machinery, engines, and city; *Sound Patterns*—man-made and natural sounds; *Sounds of My City*—sirens, subway trains, ethnic songs; *Sounds of Animals*—lions, hens, elephants, and more.

Wall charts: *Keep America Beautiful*—color photograph of Vermont autumn with Bliss Carman's poem "A Vagabond Song" (grades four to six); *Rx for Earth*—color photographs with an ecological theme (grades four to six); *Earth Life Poster*—symbolic representation of man superimposed on the world that people are destroying (junior high).

Singer SVE, 1345 Diversey Parkway, Chicago, Ill. 60614.

America's Urban Crisis, 6 sound filmstrips with records or cassettes; topics: *Air Pollution*, *Water Pollution*, *Solid Wastes*, *Transportation*, *Housing*, and *Roots of Our Urban Problems*. Upper elementary and junior high.

Communities in Nature: Ecology Learning Module, 8 filmstrips with cassettes, 4 murals 59 x 21 inches, two board games, a student booklet, a teacher's manual; theme: that a community is comprised of living things that are interrelated. Primary grade children.

Spoken Arts, 310 North Ave., New Rochelle, N.Y. 10801.

Dash McTrash and the Pollution Solution, 5 color filmstrips with cassettes; story of Dash McTrash, his friends, and their polluted town. Lower elementary.

Time-Life Inc., Time-Life Bldg., New York, N.Y. 10020.

Our Mountains of Trash, 3 sound filmstrips with cassettes and records, program guide, Life nature library book *Ecology*. Junior high and up.

Waterways or Sewers, 4 sound filmstrips with cassettes and records. program guide, Life nature library book *Water*. Junior high and up.

Troll Associates, 320 Route 17, Mahwah, N.J. 07430.

What Are Ecosystems? 6 filmstrips with cassettes, color photographs without captions, a brief teacher's guide; topics: *Pond Ecosystem, Stream Ecosystem, Salt Marsh and Seashore Ecosystem, Forest Ecosystem, Human Urban Ecosystem,* and *Comparing the City to Natural Ecosystems.* Upper elementary.

Junk Ecology, 6 filmstrips with cassettes; color photographs show children carrying out clear step-by-step instructions for converting discarded household items into practical or decorative objects; topics: *Recycling Paper and Cardboard, Recycling Tin Cans and Bottle Caps, Recycling Wood and Sticks, Recycling Plastic Throwaways, Recycling Jars and Bottles, Fix Up, Clean Up: Special Crafts.* Upper elementary.

Weston Woods, Weston, Conn. 06880.

The Last Free Bird, recording of A. Harris Stone's book that tells the story of the last bird left after man has heedlessly destroyed the birds' environment.

SCIENTIFIC SUPPLY AND EQUIPMENT COMPANIES

(A sampling of sources, large and small, comprehensive and specialized. A local high school or college science department can probably suggest additional regional sources.)

W. Atlee Burpee Co., P. O. Box 6929, Philadelphia, Pa. 19132.

Good selection of seeds and seed starting aids: peat pellets, peat pots, and heated table-top greenhouses that make raising seedlings simple.

Carolina Biological Supply Co., Burlington, N.C. 27215 and Gladstone, Oreg. 97027.

Living and preserved plant and animal materials, microscope slides, aquaria, models, charts, filmstrips, chemicals, supplies, and games such as *Dirty Water, Litterbug, Ecology,* and *Pollution.* Individual orders $10 minimum plus shipping. No chemicals sold to individuals.

Edmund Scientific Co., 42 Edscorp Bldg., Barrington, N.J. 08007.

Lenses, magnets, motors; a wide selection of gadgets and optical and general science items. Minimum order to individuals $5. Free catalog.

Fisher Scientific Co., 711 Forbes Ave., Pittsburgh, Pa. 15219.

Complete stock branches in Atlanta, St. Louis, Springfield, N.J., Washington, D.C., and other United States cities plus Edmonton, Ottawa, Quebec, and other Canadian cities. Equipment, glass and plastic ware, chemicals, and supplies.

Frey Scientific Co., 465 S. Diamond St., Mansfield, Ohio 44903.

Materials for curriculum programs—IPS, ISCS, and PSNS. Wide selection of miscellaneous items.

Macmillan Science Co., Inc. (Turtox/Cambosco), 8200 South Hoyne Ave., Chicago, Ill. 60620 and 342 Western Ave., Boston, Mass. 02135.

Complete line of science materials in biology, physical science, chemistry, science curriculum programs, and audio-visual materials. $25 minimum order from individuals.

Park Seed Co., Greenwood, S.C. 29647.

Flower and vegetable seed supplier that also has many aids for growing seeds such as seed starter kits, trays, Jiffy pots and strips, soil test kits, and garden nutrients.

Sargent-Welch Scientific Co., 7300 North Linder Ave., Skokie, Ill. 60076.
Complete stock branches in Anaheim, Dallas, Denver, Springfield, N.J., and other United States cities plus Montreal, Toronto, and Vancouver in Canada. Equipment, glass and plastic ware, chemicals, and supplies.

Stansi Educational Materials, a Division of Fisher Scientific Co., 1259 N. Wood St., Chicago, Ill. 60622.
Equipment and supplies for all sciences, science curriculum programs, and AV materials.

Ward's Natural Science Establishment, Inc., P. O. Box 1712, Rochester, N.Y. 14603 and P. O. Box 1749, Monterey, Calif. 93940.
Biology and earth science equipment and supplies, AV materials, and large collection of minerals and fossils. $15 minimum order from individuals.

SOURCES OF FREE AND INEXPENSIVE MATERIALS
(Allow at least six weeks for delivery)

Aluminum Association, 750 Third Ave., New York, N.Y. 10017.
Recycling: An Ecology Study, 80 frames, color-sound filmstrip with either record or cassette; methods of recycling solid wastes and nature's method of reusing resources. Grade six and up.

American Lung Association (National Tuberculosis and Respiratory Disease Association), regional offices throughout the United States.
Air Pollution Primer, a fact-filled book with diagrams and glossary, upper elementary and junior high. *Fact Sheets* on health effects of air pollution. Leaflets such as *Air Pollution Explained: The Pollutants*. Reprints including *Air Pollution Episodes*. Quiz leaflet "What's Your Air Pollution I.Q.?"

Atomic Energy Commission, Technical Information Center, P. O. Box 62, Oak Ridge, Tenn. 37830.
Over 50 booklets in the "World of the Atom" and "Understanding the Atom" series such as WAS-204 *Nuclear Power and the Environment*, WAS-507 *Nuclear Reactors*, and UAS-006 *Nuclear*

Terms, A Glossary. Schools and libraries may request a complete set without charge if request is made on official stationery. Send for an order form.

Classroom Science Films catalog lists free loan films available from film libraries at Oak Ridge, University of Alaska, University of Hawaii, and Puerto Rico Nuclear Center, San Juan.

The Conservation Foundation, 1717 Massachusetts Ave., N.W., Washington, D.C. 20036.

Low cost booklets on environmental quality studies; some deal with specific areas such as Galveston and San Francisco Bays, Naples, Fla., and the National Parks. Ask for the publications list.

Energy Conservation Research, 9 Birch Rd., Malvern, Pa. 19355.

The Energy Challenge—What Can We Do? a source booklet explaining what energy is, how we consume it, and listing many practical suggestions for saving energy in the home.

Forest Service, U. S. Dept. of Agriculture, Washington, D.C. 20250.

Pamphlet FS-28 *Materials to Help Teach Forest Conservation* lists bulletins, charts, posters such as AIB-305 "Your Water Supply and Forests" and PA-837 "Teaching Conservation Through Outdoor Education Activities"; bookmarks of Smoky Bear and Woodsy Owl. Pamphlet FS-31 *Forest Service Films* lists sound-color films available on free loan from regional film libraries of the Forest Service.

Keep America Beautiful, Inc., Program Development, 99 Park Ave., New York, N.Y. 10016.

Your Community Is an Environmental Resource, a pamphlet to help teachers, students, and concerned citizens treat their community as a living lab in which to observe and apply sound conservation principles; single copy free.

KAB Reports, newsletter distributed periodically that tells about KAB activities.

Litter Prevention: A First Step to Improving the Environment, a pamphlet that gives ideas for the classroom, schoolwide projects, and community action. Other booklets describe poster and essay contests, KAB/Kodak photo awards, and community action kits including litter bags.

McDonald's Corporation, One McDonald's Plaza, Oak Brook, Ill. 60525.

Ecology Action Pack, a booklet developed by the Dayton, Ohio, Museum of Natural History and containing ecology activities, spirit duplicating masters, a transparency, and cutouts for elementary grades four to six.

National Coal Association, Coal Building, 1130 17th St., N.W., Washington, D.C. 20036.

Coal in Today's World, an illustrated booklet with short paragraphs on topics including how coal was formed, kinds of coal, kinds of mines; map of coal areas in the United States shows locations and types and lists reserves by state.

National Science Teachers Association, 1201 Sixteenth St., N.W., Washington, D.C. 20036.

Environmental Education Materials Catalog (# 471–14650), a review and evaluation of about fifty free and low cost slide sets, filmstrips, films, and booklets available from government and industrial sources.

NSTA regular membership includes subscription to *The Science Teacher;* elementary membership includes subscription to *Science and Children.*

Soil Conservation Service, U. S. Department of Agriculture, regional offices listed in the telephone directory under "United States Government."

Numerous publications on soil- and water-related topics including PA-268 *An Outline for Teaching Conservation in Elementary Schools,* PA-34 *Teaching Soil and Water Conservation,* Bulletin 325 *Sediment,* and Bulletin 326 *Conservation and the Water Cycle*—a full-color poster.

TVA Environmental Education Section, 300 Farragut Hotel, Knoxville, Tenn. 37902.

Information on environmental education projects and assistance for program development available to schools of the Tennessee Valley Region. A catalog of films and sound filmstrips available at no rental charge.

TVA Information Office, 333 New Sprankle Bldg., Knoxville, Tenn. 37902.

General information on TVA programs including a listing of environmental education source materials available. A *Quality Environment in the Tennessee Valley*, an illustrated booklet details what TVA is doing in restoring fields and forests, water quality improvement, fish and wildlife, power supply, and the environment.

U.S. Environmental Protection Agency, Office of Public Affairs, Washington, D.C. 20460.

Selected Publications on the Environment lists publications available from the agency such as *Popeye and Environmental Careers*—a colorful comic book for the upper elementary crowd, *Noise Pollution*—a clear discussion of aspects of noise pollution for elementary and junior high students, and *Common Environmental Terms: A Glossary*—definitions of terms.